Anthropogenic Environmental Hazards

Pankaj Pathak • Rajiv Ranjan Srivastava •
Sadia Ilyas
Editors

Anthropogenic Environmental Hazards

Compensation and Mitigation

Springer

Editors
Pankaj Pathak
Resource Management Lab, Department of Environmental Science & Engineering
SRM University Andhra Pradesh
Guntur, India

Rajiv Ranjan Srivastava
Center for Advanced Chemistry, Institute of Research and Development
Duy Tan University
Da Nang, Vietnam

Resource Management
Faculty of Natural Sciences
Duy Tan University
Da Nang, Vietnam

Sadia Ilyas
Department of Earth Resources & Environmental Engineering
Hanyang University
Seoul, Korea (Republic of)

ISBN 978-3-031-41012-3 ISBN 978-3-031-41013-0 (eBook)
https://doi.org/10.1007/978-3-031-41013-0

This Springer imprint is published by the registered company Springer Nature Switzerland AG
The registered company address is: Gewerbestrasse 11, 6330 Cham, Switzerland

Paper in this product is recyclable.

Preface

This book exclusively focuses on anthropogenic activities responsible for environmental hazards, their compensation, and possible mitigation. The book highlights the recent anthropogenic activities which have caused major damage to the environment and led to climate change issues. With rapid demand for resources and expandable habitats causing great damage to the environment, new prospects in dealing with natural resources and also introspection on the human actions are needed, will be explained in this book. The existing anthropogenic environmental hazards will be discussed in detail, along with the environmental compensation and proper approach to mitigate environmental issues. This book is a collation of invited chapters from experts who are working in the identified areas of mining and disaster management, policymakers, and those in the environmental mitigation field.

This book is a resource material for scholars, researchers, academicians, and policymakers who are taking action to build environmental sustainability and ensuring a sustainable future. Strata of topics have been included with an objective to cover the maximum aspects of future energy mix.

The noteworthy structures of this book are associated with environmental hazards, the pros and cons of resources, environmental compensation, and mitigation to recover resources without harming the environment. Chapters 1 and 2 deals with assessing the environmental hazards due to metal mining and the transformation needs of urban mining and resource recovery. Chapters 3 and 4 emphasize metal recovery through secondary resources to compensate for environmental hazards supported by government policies. Chapters 5 and 6 focus on hazards from industrial waste and recovered metals from phytoremediation techniques. Chapter 7 gives attention to environmental sustainability while using industrial by-products such as flyash to reduce environmental burden and generate resource materials. Chapters 8 and 9 discuss mismanagement and sustainable management of municipal solid waste to bring environmental and economic sustainability. Chapter 10 describes environmental hazards and their mitigation due to biomass waste. Furthermore, collaborative governance and non-monetary compensation mechanisms for sustainable forest management and fire mitigation are discussed in Chap. 11.

Nevertheless, all the chapters in this volume will convey deep knowledge.

On the whole, the material composed in this book will bring in-depth understanding and extension of knowledge of the environmental hazards and their associated mitigation done in the recent past. Dr. Pankaj Pathak, Dr. Rajiv R. Srivastava, and Dr. Sadia Ilyas individually acknowledge the authors and reviewers for contributing to their valuable time, knowledge sharing, and their interests to bring this book into the present shape.

Guntur, India Pankaj Pathak
Da Nang, Vietnam Rajiv Ranjan Srivastava
Seoul, South Korea Sadia Ilyas

Contents

About the Editors

Pankaj Pathak is Associate Professor in the Department of Environmental Science and Engineering at SRM University, AP, India. She obtained her Ph.D. degree in Environmental Geotechnology from the Indian Institute of Technology Bombay, India. Her research domain includes solid and hazardous waste management and waste to energy, along with a keen interest in sustainable green energy resources. She has published several peer-reviewed articles in high-impact journals and edited books with ACS, Springer, CRC, etc.

Rajiv Ranjan Srivastava is an academician and industrial researcher in the areas of hydrometallurgy, resource management, and reaction thermodynamics. He earned his Ph.D. in 2017 from the University of Science and Technology (South Korea) in the major of Resources Recycling, and currently, he is Senior Lecturer at the Faculty of Natural Sciences, Duy Tan University (Da Nang), and Senior Researcher at the Center of Advanced Chemistry, Institute of Research and Development (Da Nang), Vietnam. Besides this, he has remained associated with ATTERO Recycling (India), Tae-Hyung Recycling (South Korea), Rubamin Limited (India), and CSIR—National Metallurgical Laboratory (India) in various capacities. His interdisciplinary research interests include solid and industrial waste management, urban mining, and bio- and hydrometallurgical exploitation of energy-critical elements for clean and alternative energy production. He has five patents, and published about 50 research articles in high-impact journals, some of which are among highly cited articles.

Sadia Ilyas is Brain Pool Scientist under the National Research Foundation of Korea (NRF) and Associate Professor (Research) in the Department of Earth Resources and Environmental Engineering at Hanyang University (Seoul), South Korea. She earned her Ph.D. in 2011 with doctoral research at the University of Agriculture Faisalabad focused on the metals-to-microbe interactions in the geo-environment and their application in the sustainable exploitation of valuable metals from the burgeoning legacy of the digital world (i.e., electronic waste). Dr. Ilyas has 10+ years of post-Ph.D. experience in both teaching and research in the interdisciplinary areas of bio- and hydrometallurgy, applied, inorganic, and environmental chemistry, sustainable (solid and industrial) waste management, extraction of critical raw minerals, etc. During her career, she also had the opportunity to work in different countries with well-known institutions worldwide, like Hanyang University, Seoul, Jeonbuk National University, Jeonju, and KIGAM, Daejeon, in South Korea; the University of Agriculture Faisalabad and the GC University of Faisalabad in Pakistan; and the Wuhan Institute of Technology in China. She has six patents, published about 100 research articles in high-impact journals, many of them highly cited, and wrote a number of book chapters and edited books.

Risk Assessment from Primary Mining of Precious Metal (Gold) and Possible Mitigation Route

Sadia Ilyas, Hyunjung Kim, Pankaj Pathak, and Rajiv Ranjan Srivastava

Abstract Mining is an important activity at the present time that causes severe environmental stress. The heavy metals (metalloids) easily released into the environment, surface water and groundwater contamination, and soil and air pollution are also potential risks to human health. Due to the stringent environmental rules and global thirst for achieving Sustainable Development Goals, human negligence cannot be affordable in the long run. It is time to recognize the need for people connected with safe and sustainable mining activities of lucrative precious metals like gold. Therefore, in this chapter, we assess the risks caused by the primary mining of gold and discuss mitigation routes involving various factors.

Keywords Precious metals · Artisanal gold mining · Amalgamation · Cyanidation · Acid mine drainage · Hazard index · Mining environmental liabilities

1 Introduction

Mining is related to the exploration of valuable geological minerals from the Earth and/or other astronomical sources that essentially cannot be grown by agriculture (Ilyas et al. 2021). For example, the exploration of coal, metals, gemstones, rock salt, oil shale, clay, petroleum, natural gas, etc. comes under mining activities. Among these, metal-related mining activities are one of the most important as they cater to

S. Ilyas · H. Kim
Department of Earth Resources and Environmental Engineering, Hanyang University, Seoul, Republic of Korea

P. Pathak
Resource Management Lab, Department of Environmental Science & Engineering, SRM University Andhra Pradesh, Guntur, India

R. R. Srivastava (✉)
Center for Advanced Chemistry, Institute of Research and Development, Duy Tan University, Da Nang, Vietnam

Resource Management, Faculty of Natural Sciences, Duy Tan University, Da Nang, Vietnam

P. Pathak et al. (eds.), *Anthropogenic Environmental Hazards*,
https://doi.org/10.1007/978-3-031-41013-0_1

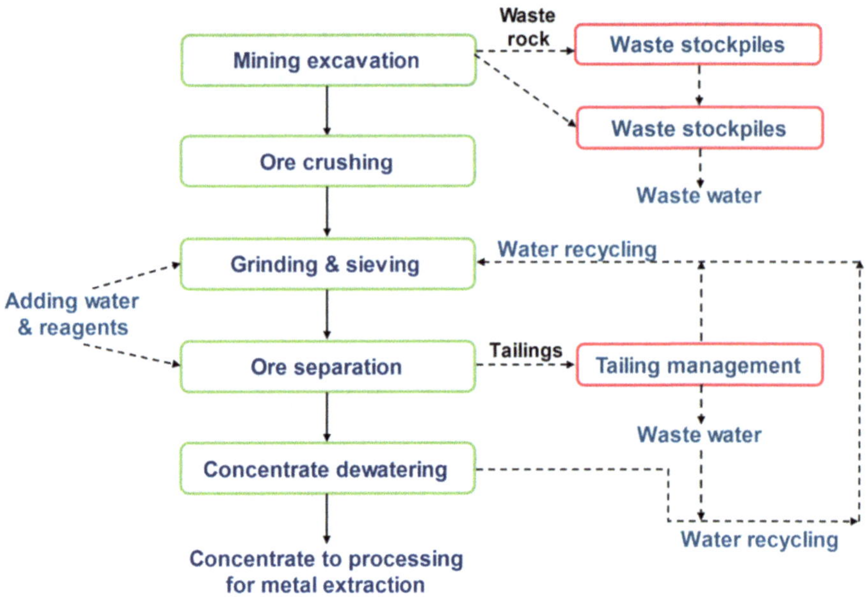

Fig. 1 Schematic representation of typical mining activities at a standard condition

our 24-hour needs (Ilyas et al. 2018). In general, the mining of minerals can be divided mainly into five stages: mining, crushing, grinding, ore concentration, and dewatering (Srivastava et al. 2023). The typical operations in any mining activities carried out at a standard condition are illustrated in Fig. 1. From these activities, mainly two types of waste are generated: waste rock piles (tailings) and wastewater. Wastewater is discharged from either the underground runoff water or the used water in grinding and ore separation, whereas tailings are produced from the grinding and ore separation processes and go to the tailing management facilities (Global Tailing Review 2020). In most cases, tailing dams are mining legacies that release potentially toxic elements (viz., Cd, As, Cu, etc.) into nearby water sources to be used in irrigation of agricultural land and sometimes uptaken by humans (Dong et al. 2020). Poor management of tailing waste has caused tailing dam failures in the past with catastrophic consequences (Grebby et al. 2021; Ouellet et al. 2022). Ideally, the wastewater should be recycled between the grinding, separation, and dewatering stages; however, it is not always easy to practice. On the other hand, the final product obtained after the mining activities, i.e., ore concentrates, is sent to the extraction and purification processes for the recovery of pure metals.

Mining activities can have both positive and negative impacts on society and the environment during the ongoing mining process or after the abandonment of a mine, both in direct and indirect ways. Exploration, construction, and operation often cause land-use change to have negative impacts on the environment due to associated issues like deforestation, soil erosion, contamination of water streams and wetlands, an increase in noise pollution due to machining, and air pollution due to dust and emissions (Cruzado-Tafur et al. 2021). Similarly, abandoned mines result in remarkable environmental footprints like soil and water contamination (Ngole-Jeme and Fantke 2017). Beyond the mining activities, the built infrastructure like roads, railways, ports, electricity, etc. can affect the flora and fauna leading toward the migratory routes (Rapant et al. 2006). Nevertheless, mining has positive impacts in terms of employment generation, not only in mining itself but also in other sectors like transportation, metallurgical and chemical industries, and mushrooming local businesses for daily needs. The concerns over environmental risks and public health issues are greater because damage cannot be compensated. Up to some extent, remediation of the potential environmental impacts by means of ecological restoration and wastewater treatment (Paniagua-López et al. 2021). As prevention is better than cure, the nations are trying to regulate the mining activities to control the damage; however, the full enforcement of regulations is always challenging.

The risk assessment of primary mining may be helpful to identify related issues, estimate the probability and severity of the consequences, and make strategies for environmental management (Cervantes Neira and Quito Quilla 2020). Accordingly, risk assessment is a process that comprises measuring the possible adversities resulting from the exposition of environmental stress, including the physical, chemical, and biological entities responsible for damaging the ecosystem (USEPA 1998). Given that mitigation measures can be implemented to avoid, eliminate, and reduce the negative or improved impacts, such measures must be outlined in environmental and social impact assessments before major mining activities (Arranz-González et al. 2021). The mitigation of environmental impacts in one system can influence other systems. For example, water or soil treatments are linked to the well-being and health of local inhabitants and biodiversity. A variety of technological advancements have been seen in the area of wastewater treatment and contaminated land (Cervantes Neira and Quito Quilla 2020; USEPA 1998). In contrast, the mitigation routes designed to alleviate negative impacts on the environment and society may not always be effective (Haddaway et al. 2019). Indeed, the risk assessment from primary mining of precious metals and possible mitigation routes have been poorly focused, as has been the aim of this chapter.

2 Risk Assessment

Risk assessment is related to the analysis and mitigation of risks in the form of natural disasters and mining accidents in terms of their impacts on the environment and society. The natural hazards and disaster risks include earthquakes, landslides, floods, etc., whereas the mining industrial hazards typically involve mine sliding, dam failures, fire and explosions, water leakage, the release of hazardous chemicals or radionuclides, etc. In order to assess the risk, the following approaches are desirable (EIA Guidelines 2018):

(i) Identifying the category of hazards and/or disasters from the historical available data
(ii) Considering the case of climate change and its implications for frequency and consequences
(iii) Estimation of spatial patterns, time, frequency, and intensity
(iv) Identification based on project design and layout, use and handling of hazardous substances from the case studies, available records, and media reports
(v) Analyzing cause and effect and the probability of events
(vi) Assessing the extent of damages by accounting for the layout and design of the project, exposure routes and environment, local inhabitants, etc.
(vii) Calculating the overall risk and comparing it with the acceptable levels
(viii) Identifying the need for mitigation measures

2.1 *Mercury Load in Fish*

Several routes exist for environmental risk assessment and modeling, along with their limitations and advantages (Kammen and Hassenzahl 2001). Determining the metal concentrations in fish and soil samples is simple to establish the safe-unsafe or maximum concentration level thresholds. Recently, Marcantonio et al. (2021) reported the mercury load in fish from a precious metal mine in Sierra Leone to be 302 ppb. Although the sampled fish were mostly short-lived freshwater species, the Hg concentration was average for albacore tuna (FDA 2018). About 87% of fish samples exceeded in Hg concentrations the FDA recommended value of 150–230 ppb for 2–3 meals/week consumption, whereas 15% of samples exceeded the zero meals/week threshold with >460 ppb Hg therein. The one-way ANOVA for Hg concentration and 95% confidence interval for the range of hazards index are shown in Fig. 2a, b, respectively.

2.2 *Heavy Metals in Soil*

The soils near the mining activities are always found to be severely contaminated with heavy metals and sometimes radionuclides as well, which pose potential risks to human health. The majority of soil samples near the primary mining of precious metals contain cobalt, chromium, lead, iron, and thorium in concentrations that exceed most established limits (as per Table 1). For example, the soils collected along the Pampana River in Sierra Leone (Africa) were analyzed to contain 115 ppm Th in comparison to the limit of 7.4 ppm (US-EPA 2019). Notably, the risk posed by heavy metals can be assessed by following the steps of the US-EPA (2014):

(i) *Hazard identification:* a process that determines whether exposure to a stressor increases the likelihood of adverse health effects

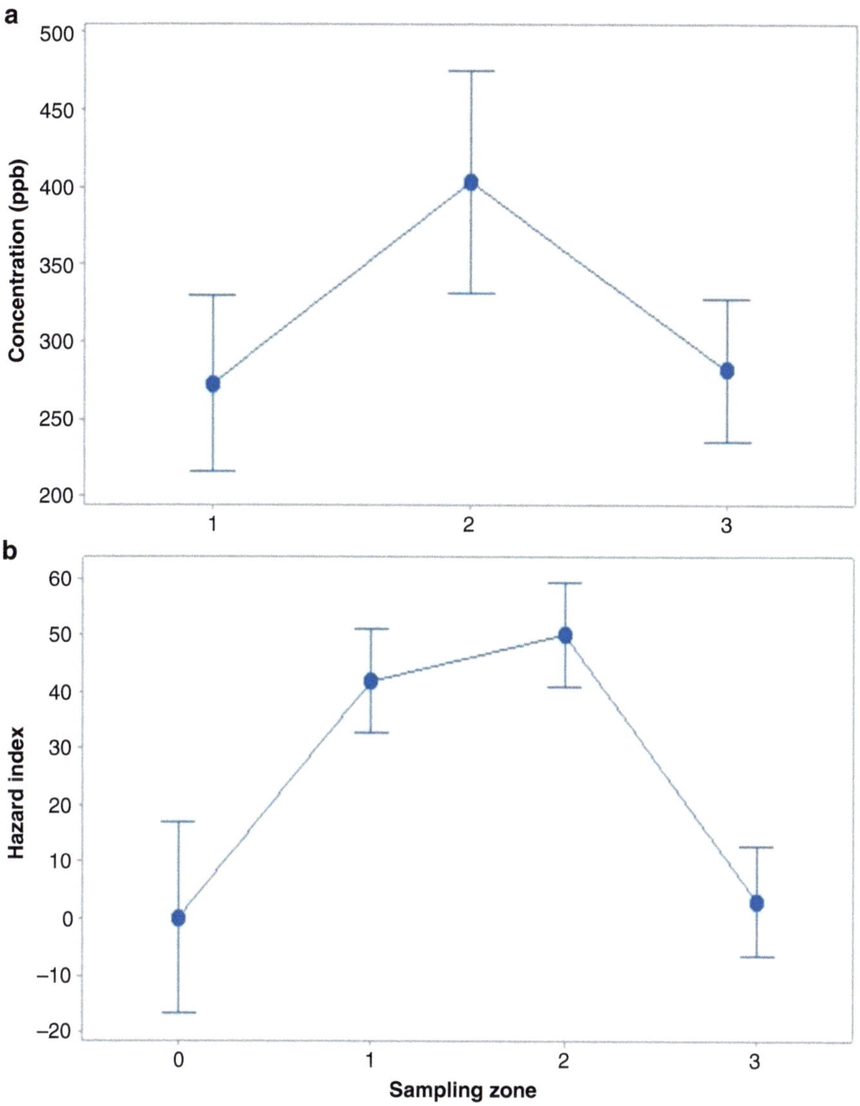

Fig. 2 Interval plots of testing for differences in Hg content in fish (**a**) and T-test results for greater probability of risk than EPA established noncarcinogenic risk acceptable risk: HI > 1 (**b**) (modified and adopted from the open supplementary source of Marcantonio et al. 2021)

(ii) *Exposure assessment:* a process that estimates the quantity of heavy metal exposure to a human being
(iii) *Dosage-response assessment:* a process that determines the human health problems associated with different uptakes of heavy metals

Table 1 Heavy metals' limits in soil samples as fixed by different countries (Atanassov 2008; Canada-MoE 2011; Kamunda et al. 2016; Li et al. 2019; Rodríguez-Eugenio et al. 2018; SA-DEA 2010; Tóth et al. 2016; UK-DoE 2009), NA = not analyzed

Heavy metals (mg/kg)	France	Belgium	Hungary	Germany	Netherlands	Finland	Poland	UK	Australia	Taiwan	Canada	South Africa	China
As	37	110	15	50	55	50	2	37	20	60	11	5.8	30
Cd	20	6	1	20	12	10	4	22	3	5	1	7.5	3
$Cr_{(total)}$	130	300	75	400	380	200	NA	130	50	250	67	6.5	1000
Cu	190	400	30	NA	190	150	150	NA	100	NA	19	300	NA
Hg	7	15	0.5	20	10	2	NA	10	1	2	0.16	0.93	2
Pb	400	700	100	400	530	200	100	200	300	300	45	20	80
Ni	140	470	40	140	210	100	NA	130	60	200	37	91	100
Zn	9000	1000	200	NA	720	250	300	NA	200	600	290	240	250
Co	NA	NA	NA	NA	NA	100	NA	NA	100	NA	19	300	NA

Table 2 Exposure parameters (SA-DEA 2010; US-EPA 2011)

Parameter	Unit	Adult	Child
BW—body weight	kg	70	15
CF—conversion factor	kg/mg	10^{-6}	10^{-6}
ABS—dermal absorption factor	n/a	0.1	0.1
FE—dermal exposure ratio	n/a	0.61	0.61
EF—frequency of exposure	days/years	350	350
DE—duration of exposure	years	30	6
IR_{air}—inhalation rate	m^3/day	20	10
IR—ingestion rate (IR)	mg/day	100	200
PEF—particulate emission factor	m^3/kg	1.3×10^9	1.3×10^9
SA—skin surface area	cm^2	5800	2100
AF—soil adherence factor	mg/cm^2	0.07	0.2
AT—average time (carcinogens)	days	365 × 70	365 × 70
AT—average time (noncarcinogens)	days	365 × 30	365 × 6

(iv) *Risk characterization:* a process that determines the extra risk of health issues for the exposed population

For the risk assessment, determining the average daily intake (ADI) is a vital factor, which is the intake value of a particular heavy metal from soil by any means of ingestion, inhalation, or dermal intake. ADIs are calculated using exposure parameters with respect to age, body size, respiration rates, etc. Hence, adults and children have different parameters, as summarized in Table 2 (SA-DEA 2010; US-EPA 2011).

Thus, ADI for three different intakes (i.e., ingestion, inhalation, and dermal intake) can be calculated using the following equations:

$$\mathrm{ADI}_{(\mathrm{ingestion})} = \frac{C \times \mathrm{IR} \times \mathrm{EF} \times \mathrm{ED} \times \mathrm{CF}}{\mathrm{BW} \times \mathrm{AT}} \tag{1}$$

$$\mathrm{ADI}_{(\mathrm{inhalation})} = \frac{C_s \times \mathrm{IR}_{\mathrm{air}} \times \mathrm{EF} \times \mathrm{ED}}{\mathrm{BW} \times \mathrm{AT} \times \mathrm{PEF}} \tag{2}$$

$$\mathrm{ADI}_{(\mathrm{dermal})} = \frac{C_s \times \mathrm{SA} \times \mathrm{FE} \times \mathrm{AF} \times \mathrm{ABS} \times \mathrm{EF} \times \mathrm{ED} \times \mathrm{CF}}{\mathrm{BW} \times \mathrm{AT}} \tag{3}$$

Further, the noncarcinogenic hazards are characterized by a unitless hazard quotient (HQ) representing the probability of individual health adversity (US-EPA 2014). It is the ratio of ADI and chronic reference dosage (RfD) of a heavy metal (given in Table 3) that can be written as below:

$$\mathrm{HQ} = \frac{\mathrm{ADI}}{\mathrm{RfD}} \tag{4}$$

Furthermore, the hazard index (HI) for *n* number of heavy metals can be determined as the sum of all HQ for a soil sample as follows:

Table 3 The chronic reference dosage (RfD) of heavy metals in mg/kg-day (SA-DEA 2010; US-EPA 2018)

Heavy metals	Ingestion RfD	Inhalation RfD	Dermal RfD
As	3.00E-04	3.00E-04	3.00E-04
Hg	3.00E-04	8.60E-05	3.00E-04
Cd	5.00E-04	5.70E-05	5.00E-04
Cr	3.00E-03	3.00E-05	–
Co	2.00E-02	5.70E-06	5.70E-06
Pb	3.60E-03	–	–
Ni	2.00E-02	–	5.60E-03
Cu	3.70E-02	–	2.40E-02
Zn	3.00E-01	–	7.50E-02

Table 4 The cancer slope factors (CSF) are in mg/kg-day (SA-DEA 2010; US-EPA 2018)

Heavy metal	Ingestion CSF	Inhalation CSF	Dermal CSF
As	1.50E+00	1.50E+01	1.50E+00
Cd	–	6.30E+00	–
Cr	5.00E-01	4.10E+01	–
Co	–	9.80E+00	–
Pb	8.50E-03	4.20E-02	

$$\mathrm{HI} = \sum_{k=1}^{n} \mathrm{HQk} = \sum_{k=1}^{n} \frac{\mathrm{ADI}}{\mathrm{RfD}} \tag{5}$$

If HI < 1, heavy metal exposure is unlikely to cause health adversities. If HI > 1, the exposure may be concerning for potential noncarcinogenic effects.

The carcinogen risk is the unitless probability of a person developing cancer from heavy metal exposure over a lifetime (US-EPA 2014). It can be determined as a total of the lifetime cancer risk of a person from the average contribution of the individual heavy metals for all the pathways as follows:

$$\mathrm{Risk}_{(\mathrm{pathways})} = \sum_{k=1}^{n} \mathrm{ADI}_k \mathrm{CSF}_k \tag{6}$$

ADI_k (mg/kg-day) is the average daily intake, and CSF_k (mg/kg-day) is the cancer slope factor, respectively, for the *k*th heavy metal and the *n* number of heavy metals. The CSF values for the incremental risk of an individual developing cancer can be seen in Table 4. A total cancer risk can finally be calculated as the sum of each individual pathway and heavy metal for a given soil sample as follows:

$$\mathrm{Risk}_{(\mathrm{total})} = \mathrm{Risk}_{(\mathrm{ingestion})} + \mathrm{Risk}_{(\mathrm{inhalation})} + \mathrm{Risk}_{(\mathrm{dermal})} \tag{7}$$

where $Risk_{(ingestion)}$, $Risk_{(inhalation)}$, and $Risk_{(dermal)}$ are the risks posed via by ingestion, inhalation, and dermal pathways, respectively.

The exposure risk quantified by Marcantonio et al. (2021) revealed a higher hazard index value for the inhabitants living near the Lake Sonfon area of gold mining. They used heavy metal concentration and the hazard index (HI) to determine the noncarcinogenic risk to adults and children separately in wet and dry seasons (refer to Fig. 3 adopted and modified from Marcantonio et al. 2021). For adults, the mean HI was 36.8 and 28.8 in the wet and dry seasons, respectively, and 67.6 and 54.1 for children. As per the US-EPA, a HI value greater than 1 represents excessive risk; hence, significantly high noncarcinogenic risks to both adults and children were observed. The results also revealed that HI risk greatly increases by going near the Lake Sonfon gold mine site. Furthermore, the evaluated values of carcinogenic risk (CR) causing cancer due to heavy metal exposure showed high risks closest to the gold mining area. Notably, the US-EPA guidelines state the threshold of 1×10^{-4} to 10^{-6}; however, the mean CR was found to be 1.01×10^{-3} and 9.42×10^{-3} for adults and children, respectively, indicating that the gold mining area is highly contaminated by heavy metal pollution with a great likelihood ratio.

2.3 *Mining Environmental Liabilities (MEL)*

On the other side, the Spanish Geological Survey proposed a simplified risk assessment of abandoned mine sites to evaluate the mining environmental liabilities (MEL) through the associated determined probability (IP) and severity indices (IS) (refer to Fig. 4). For mine waste deposits, the risk assessment was performed by adopting the protocol designed by the Spanish Geological Survey (Alberruche del Campo et al. 2014), whereas in the case of other (i.e., non-waste) deposits, the methodology outlined in the Environmental Risk Assessment Guide of the Ministry of the Environment of Peru was followed (MINAM 2010). The cartographic information was processed using Geographic Information Systems (GIS) through ArcMap 10.8.1 to identify the areas that represent a greater risk of affecting the flora and fauna. Such information gathering allows for better management actions in each evaluated area of concern.

In the recent past, gold mines have gone deeper, encountering arsenopyrite-bearing gold ores that inhibit the leaching process during gold extraction. The excavated ore undergoes amalgamation or cyanidation after the fine milling operation, which produces a large quantity of tailings. These tailings contain unbroken pyrites, which may lead to slow atmospheric oxidation (Naicker et al. 2003). Near Johannesburg in South Africa, the long exposure to oxygenated rainwater with the undisturbed tailings caused the oxidation of pyrite minerals up to 5 m below the soil surface (Marsden 1986). The sulfate produced via pyrite oxidation acidifies the groundwater and enters streams along with Witwatersrand, severely polluting the groundwater and soils. The gold-tailing impoundment in the Witwatersrand can

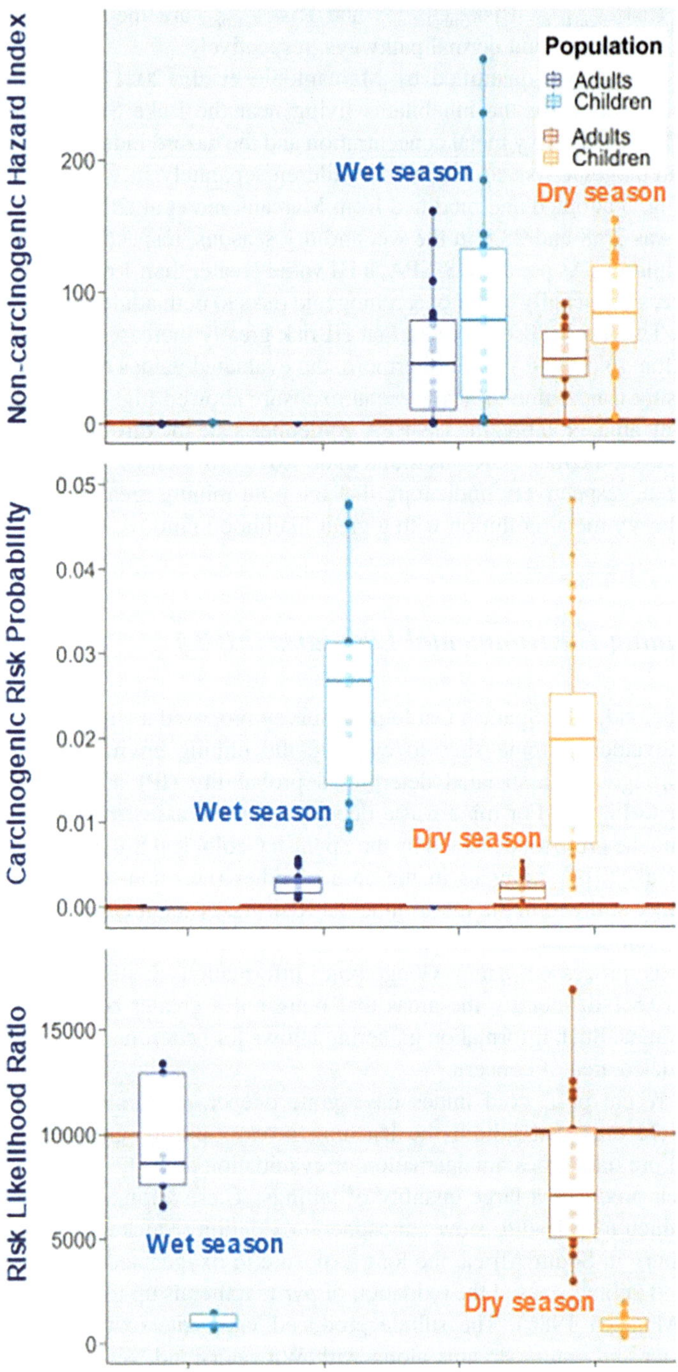

Fig. 3 Noncarcinogenic hazard index, risk probabilities, and likelihood ratio evaluated at two seasons of wet and dry and determined for adults and children (red line showing the threshold lines as described by Marcantonio et al. 2021)

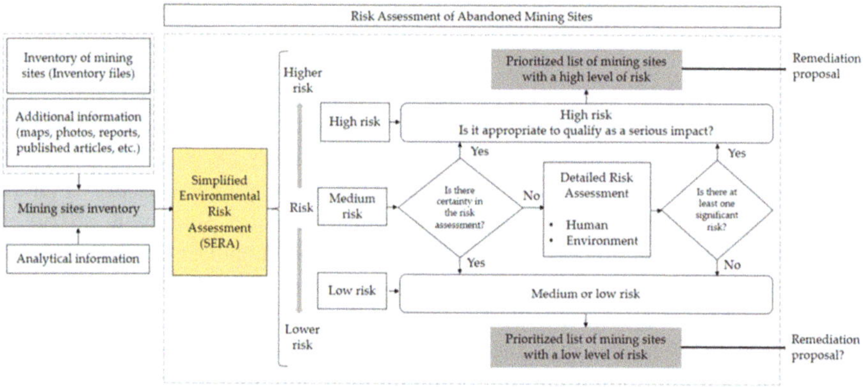

Fig. 4 Simplified schematic for the risk assessment of abandoned mining sites as designed by the Spanish Geological Survey (adopted from the open access source of Salgado-Almeida et al. 2022, MDPI)

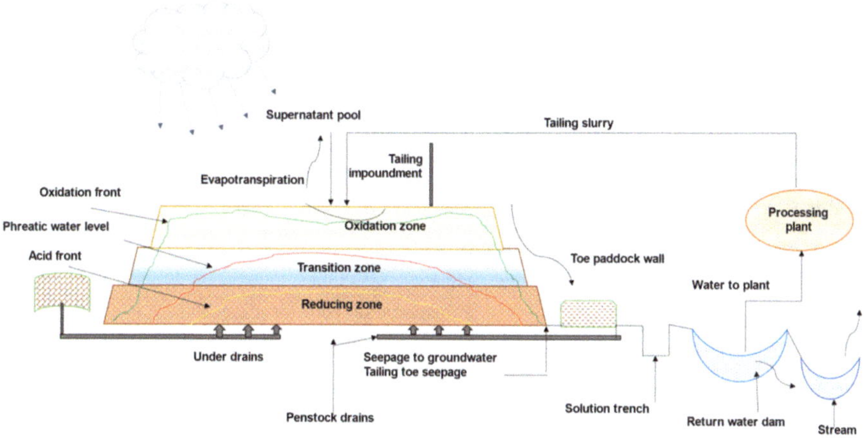

Fig. 5 A schematic of a typical gold-tailing impoundment in the Witwatersrand (modified and adopted from Hansen 2015 with copyright permission from Elsevier)

typically be understood by the scheme presented in Fig. 5 (Hansen 2015). Naicker et al. (2003) found that acid contamination in soil and groundwater varies seasonally. They found that the pH of the water near the mining area remains about 5.0, along with high concentrations of heavy metals and sulfate ions. This also leads to the precipitation of calcium as gypsum along the river banks.

3 Mitigation of Environmental Risks

In order to avoid and/or reduce environmental impacts, designing mitigation measures is an imperative action. In the case of non-avoidable environmental impacts, the restoration and rehabilitation plan should be considered on a priority basis. Additionally, the impacts should be offset through programs to improve lands, water streams, and/or facilities to offset biodiversity impacts, along with providing the facilities or compensation to offset the societal and economic impacts. Table 5 indicates the mitigation hierarchy for handling the environmental risks due to mining activities (EIA Guidelines 2018). As prevention is always better than expensive remediation, the proactive avoidance of negative impacts with precautionary steps is welcomed when the environmental consequences of mining activities can't be predicted and thus reliably managed.

A risk assessment study of MEL for their categorization and prioritization in gold-mining areas of Macuchi, Tenguel-Ponce Enriquez, and the Puyango River Basin in Ecuador, conducted by Salgado-Almeida et al. (2022), revealed that the impacts are mainly associated with artisanal and small-scale gold mining. Lack of regulations in many illegal mining activities (Rivera-Parra et al. 2021) has caused MEL accumulation. The same accumulation has facilitated the transportation of pollutants in different environmental compartments to spread severe anthropogenic contamination (SENAGUA 2011).

Table 5 Mitigation hierarchy for mining activity

Measures	Hierarchy	Actions
Avoid		• Mine site selection
		• Transportation corridor alignments
		• Mining layout for the facilities
Reduce		• Minimize the pollution and waste generation
		• Reduce the land take and disturbance
		• Limiting the use of water and energy
Restore and rehabilitate		• Exploration drilling sites
		• Rehabilitation of disturbed areas
		• Restoration of abandoned mines (vegetation, trees, wildlife, etc.)
Offset		• Provide facilities to offset societal and economic impacts
Enhance		• Community development programs for social benefits

Table 6 Categorization of priority area and proposed strategies for the pollution control under the mining environmental liabilities (MEL)

MEL	Proposed actions
Landfilling	Covering, sealing, and revegetation of deposits; physiochemical stability control and monitoring plan; water, soil, sediment, biotic component, and stability control
Mining galleries	Physiochemical stability control and monitoring plan; implementation of geoparks and museums, plugging of higher-risk mine entrances/galleries
Mine entrances	
Tailing deposits	Resource utilization of mine tailings; covering, sealing, and revegetation plan near the tailing ponds; water, soil, sediment, biotic component, and stability control
Abandoned infrastructure	Construction of a community meeting place; water, soil, sediment, biotic component, and stability control
Mineral processing plants	Control and monitoring of chemical stabilization of soils; dismantling infrastructures
Alluvial terrace	Restitution of flora and fauna; treatment of water bodies; physiochemical stabilization of tailings/riverbanks
Quarries	Physiochemical stability of soils; revegetation; water, soil, sediment, biotic component, and stability control

Table 6 belongs to the strategies and actions proposed for MEL management to mitigate the high risk for the flora and fauna (Salgado-Almeida et al. 2022). Puyango and Tenguel-Ponce Enriquez have been identified as priority control areas and urgently require a solid restoration and rehabilitation plan. To ensure the ecological restoration of the abandoned mine sites, actions like phytoremediation (Lam et al. 2017; Vela-García et al. 2019) and the creation of geoparks in low-risk areas (Franco et al. 2020), along with a continuous control and monitoring plan, could ensure the identical recovery and restitution of the land. In the entire restoration process, the collaborative efforts of different parties that involve mining companies, mine planners, investing organizations, local bodies, and societies (Popovic et al. 2015 are greatly required to achieve the Sustainable Development Goals. Lack of which may lead to an inefficient output. For a sustainable handling of mine tailings, their reutilization in metal recovery and reuse in other forms, like in the construction industry to fabricate bricks, cement, and ceramic materials, can lead toward a circular economy (Srivastava et al. 2023). Finally, a risk communication plan must exist to reduce exposure to potentially hazardous materials.

In order to respect human rights as a fundamental operating principle for mine workers and other actors as well, a typical compliance program can be designed as depicted in Fig. 6. In accordance with the Sustainable Development Goals, leading from the top and embedded throughout the organization, shared learning, partnership, and collaboration are the key factors to mitigate the negative impacts and ensure close monitoring of the mining activities.

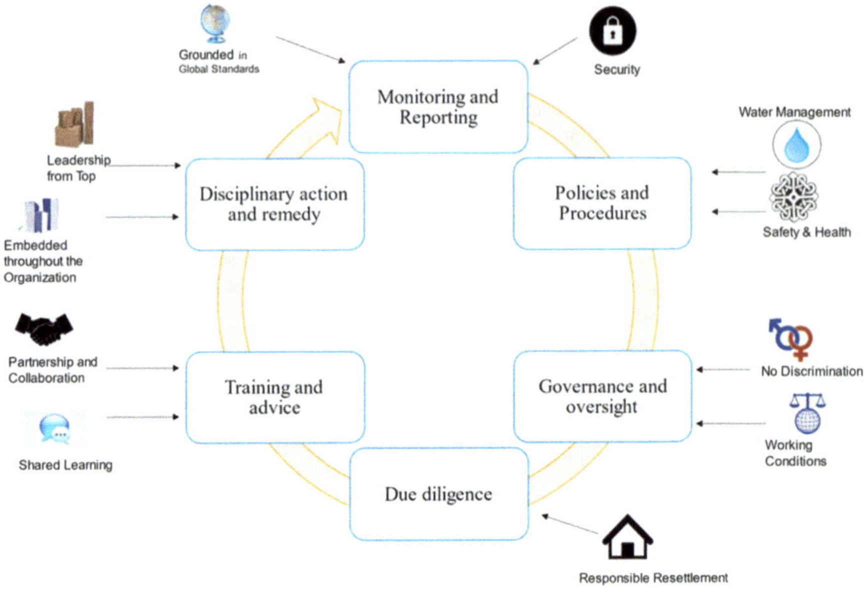

Fig. 6 The compliance program based on the Barrick Gold practicing

4 Mitigation of Water Pollution Risks

In addition to the large-scale mines, about 15%–20% of gold is extracted by artisanal and small-scale miners without any facilities or mechanical supports. Due to their informal activities, they often do not follow safety and environmental regulations, mainly resulting in cyanide and mercury pollution of water bodies and soils (Ilyas and Lee 2018). Even the highly toxic cyanidation is an industrial process of gold production, whereas the artisanal miners use mercury for dissolving the gold traces from the collected mud, soil, and rocks. As an estimate, about 140 kg of cyanide and a massive consumption of 700 tons of water are required to produce 1 kg of gold (Mudd 2007). The leached cake or slurry often contains waste rocks with heavy metals like arsenic and copper (biodegradation paper), which are stored either in open dumping or within a dam. On the other side, it is estimated that more than 2000 tons of mercury have been released into the Amazon River since the 1980s from mining activities (Malm 1998). Porcella et al. (1997) additionally claim to release 460 tons of mercury per year from small-scale mining alone. Arsenic is another toxicant of greatest concern because of its carcinogenic potential (RAIS 2021). As concentration in freshwater varies in the range of 0.15–0.45 μg/L (Singh et al. 2015), which can exceed in mining environments (Guzmán-Martínez et al. 2020). For example, the Central Andes region of Peru contains 14–23 μg/L of arsenic (Custodio et al. 2020), artisanal gold mine areas of Colombia contain 0.6–52 μg/L arsenic (Alonso et al. 2020), and mine sites in Slovakia contain 0.5–103 μg/L arsenic (Rapant et al. 2006).

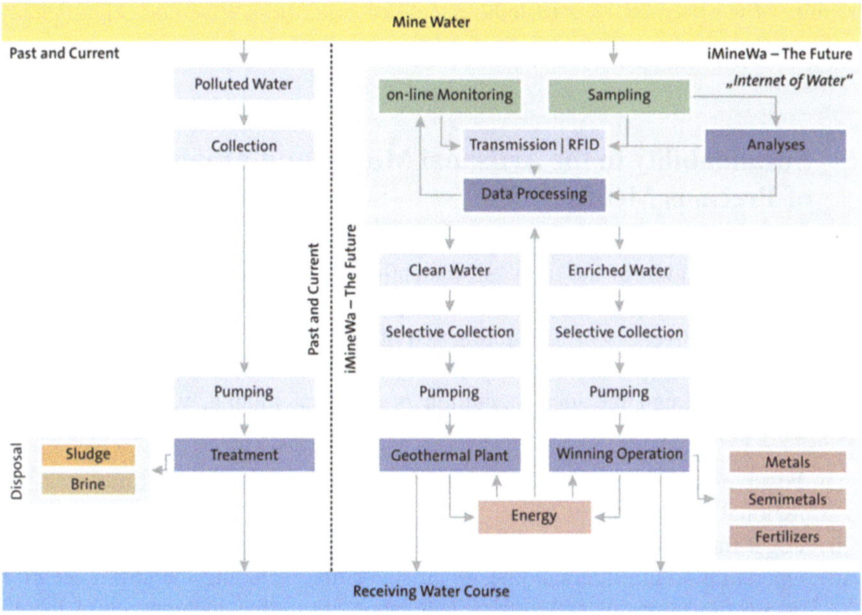

Fig. 7 Comparing the past and current mine water treatment practices with the future proposed digitized management of mine water (modified and adopted from More et al. 2020 with copyright permission from Elsevier)

Therefore, an effective treatment of mine water, especially in abandoned mines, is desperately sought. However, the current practices of mine water treatment are not ideal as they are based on the composition and volumes of water entering the plant (refer to Fig. 7a). It means that the treatment plant needs to react instantly whenever the volumes or chemistry of mine water change (More et al. 2020). Usually, no interaction is observed between water inflow into the mine, technological changes within the mine, water analyses, and outflow of treated water (More and Wolkersdorfer 2019). Here, the application of digital technologies can have a timely response by interacting with the factors involved, proving the need for a technological solution (as shown in Fig. 7b).

The polluting effect of acid mine drainage is particularly pronounced in the upper catchments of the Blesbokspruit and Klip Rivers in South Africa, which drain the southern Witwatersrand escarpment (Ilyas and Lee 2018). The water discharge of the accumulated volume in the voids of closed gold mines on the Witwatersrand to neighboring mines, generally of low quality, necessitates basic additional treatment by lime to raise the pH and air blowing to oxidize and precipitate the iron and other heavy metals. The iron was allowed to settle and disposed of on tailing dumps while discharging the water into local rivers. Although the water discharge has a neutral pH, it also has a very high sulfate (1500 mg/L) concentration and thus adds more pollution to the load already carried by the rivers in mining areas. The pollution arising from gold mines in the Central and Western Basins is well illustrated by the salinity of the Vaal River; a periodic release of water from Vaal Dam significantly reduces the salinity for downstream Vaal River users.

5 Sustainability in the Artisanal Mining and Amalgamation of Precious Metals

As per the definition of the term "sustainability," it depends upon the factors like society, environment (including socio-ecological factors), and the economy (related to the cost-effectiveness of the mining activities). Despite the issues in defining a sustainable mining operation, the experiences from the past and current activities can be helpful to design the ideal condition of artisanal mining, which include the following points of:

(i) Positive contribution of ASGM activities in development of rural area and regional empowerment
(ii) Legal framework in harmony with the national mining sector policies
(iii) Operation within international social standards, including the social security, occupational health and safety, and labor laws (that includes child labor), education and medical facilities, etc.
(iv) Environmentally sound operation with scientific and mechanized inputs
(v) Harmony between the small operations and large-scale mining operations
(vi) Ensuring high recovery yield, including a systematic development of the deposits and continuous operation

5.1 Technical Aspects

It has been observed that many problems existing with the artisanal mining can be resolved by the use of appropriate technical solutions. A prominent example can be of mercury emissions in an artisanal practice which can be resolved by using the end-of-pipe technology (involving filter, retort, and trap system). Technological issues often require technical solutions albeit an integral approach is crucial. In contrast to the traditional use of low and non-mechanized activities, the design of new milling and alternative to simple stone mortar amalgamation mills does not involve very high level of technological understanding. Hence, the conventional mining equipment are frequently modified and maneuvered by the miners to fulfill their demands for the high throughput and efficiency; however, in most cases, the suppression of security features are very unfortunate (e.g., water supply for drilling hammers). It is noteworthy that although it remained in practice since several centuries, it has been labeled as an unorganized practice, which makes it imperative to pay attention by the researchers and environmentalists as well.

5.2 *Policy and Legal Framework*

To achieve sustainability in artisanal mining of precious metals via integrating the rural development with the associated economic benefits, a policy framework development is desirable that can be based on the following strategies (Ilyas and Lee 2018):

(i) Poverty alleviation
(ii) Optimization of the business climate for the small mining sector
(iii) Insurance of sustainability
(iv) Stabilization of government revenues from the sector

Numerous reasons exist for the continuation of artisanal mining within the informal sector, majorly because of less understanding on the legal requirements. Lacking capacity to enforce penalties and to provide benefits, which should be associated with legalization, acts as a further disincentive to miners to be legalized. By recognizing the capability and contribution of the sector in the precious metal mining and metallurgy, the governments need to develop a consistent and holistic approach. In the recent times, the reforms in national policy has initiated the drive toward enabling a legal framework in the countries like Colombia, Peru, South Africa, and Tanzania, mainly beneficial to the sustainable management and exploitation of mineral resources along with the promotion of investment and licensing of the artisanal mining (Ilyas and Lee 2018). Additionally, the organization of this sector as a community or society should be promoted by the local administrations to formalize the informal structures via coordination and a harmonized management of the natural resources.

Currently, in many countries, the mining laws or other legal instruments do not support the development of small industries based on local mining production. This is especially valid for the production of informal artisanal mining activities, which is difficult to integrate into the formal economy. Training resources for healthcare providers that directly address artisanal- and amalgamation-related health issues are scarce. However, case studies, toxicology, and occupational health literature and publications from governmental and nongovernmental organizations do contain or suggest health components that could be developed further for use in this context. By using the principles of fair-trading, small-scale producers in developing countries are given the opportunity to trade their products under better selling terms and conditions. An improved awareness on health hazards is needed to practice a better, healthier, eco-friendly, and sustainable technologies in the mining of precious metals.

6 Conclusions

Looking at the importance of mining, despite the fact that it poses a threat to the environment and human health, this chapter provides a preliminary assessment of the risks associated with the present mining activities of gold involving artisanal, small-scale, and industrial gold mining. It has been found that heavy metals (metalloids) are easily released into the environment through surface water and groundwater contamination and soil and air pollution. Therefore, it is necessary to remediate the polluted mine sites and practice continuous monitoring to restore the ecosystems properly. In addition, the population's exposure must be restricted to high-risk areas through a communication plan about the risks. The mitigation hierarchy for mining activities is discussed to avoid, reduce, restore and rehabilitate, offset, and enhance communities, along with the proposed strategies for pollution control. For the potential mitigation of the risks, digitized mine water treatment and a compliance program for mine workers have been suggested to achieve the Sustainable Development Goals in the field of primary gold mining.

Acknowledgments The authors SI and HK acknowledge their contributions through the Basic Science Research Program, National Research Foundation of Korea (NRF), funded by the Ministry of Education (Project no. 2023-00243477), and grant funded by the Korea Government (MSIT) (No. 2022R1A5A1032539).

References

Alberruche del Campo ME, Arranz-González JC, Rodríguez-Pacheco R (2014) Manual Para La Evaluación de Riesgos de Instalaciones de Residuos de Industrias Extractivas Cerradas o Abandonadas, IGME

Alonso DL, Pérez R, Okio CK, Castillo E (2020) Assessment of mining activity on arsenic contamination in surface water and sediments in southwestern area of Santurbán paramo, Colombia. J Environ Manage 264:110478

Arranz-González JC, Rodríguez-Gómez V, Fernández-Naranjo FJ, Vadillo-Fernández L (2021) Assessment of the pollution potential of a special case of abandoned sulfide tailings impoundment in Riotinto mining district (SW Spain). Environ Sci Pollut Res 28:14054–14067

Atanassov I (2008) New Bulgarian soil pollution standards. Soil Chemical Pollution, Risk Assessment, Remediation and Security. Springer, Cham, pp 129–138

Canada-MoE (2011) Soil, ground water and sediment standards for use under Part XV.1 of the Environmental Protection Act. Ministry of Environment and Climate Change

Cervantes Neira JJ, Quito Quilla SJ (2020) Evaluación de riesgo ambiental generado por pasivo ambiental minero en la calidad de agua superficial. Natura@economía 5(1):1–14

Cruzado-Tafur E, Torró L, Bierla K, Szpunar J, Tauler E (2021) Heavy metal contents in soils and native flora inventory at mining environmental liabilities in the Peruvian Andes. J South Am Earth Sci 106:103107

Custodio M, Cuadrado W, Peñaloza R, Montalvo R, Ochoa S, Quispe J (2020, 1946) Human risk from exposure to heavy metals and arsenic in water from rivers with mining influence in the Central Andes of Peru. Water 12(7)

Dong L, Deng S, Wang F (2020) Some developments and new insights for environmental sustainability and disaster control of tailings dam. J Clean Prod 269:122270

EIA Guidelines (2018) Environmental Impact Assessment Guidelines for the Mining Sector. Prepared by Myanmar Mining EIA Guidelines Working Group with the technical assistance of ADB TA 8786-MYA: Environmental Safeguard Institutional Strengthening

FDA (2018) Mercury levels in commercial fish and shellfish (1990–2012). US Food and Drug Administration, Washington, DC

Franco GH, Mero PC, Carballo FM, Narváez GH, Bitar JB, Torrens RB (2020) Strategies for the development of the value of the mining-industrial heritage of the Zaruma-Portovelo, Ecuador, in the context of a Geopark project. Int J Energy Prod Manag 5:48–59

Global Tailing Review (2020) Global industry standard on tailings management

Grebby S, Sowter A, Gluyas J, Toll D, Gee D, Athab A, Girindran R (2021) Advanced analysis of satellite data reveals ground deformation precursors to the Brumadinho Tailings Dam collapse. Communications Earth & Environment 2(1):2

Guzmán-Martínez F, Arranz-González JC, Ortega MF, García-Martínez MJ, Rodríguez-Gómez V (2020) A new ranking scale for assessing leaching potential pollution from abandoned mining wastes based on the Mexican official leaching test. J Environ Manage 273:111139

Haddaway NR, Cooke SJ, Lesser P, Macura B, Nilsson AE, Taylor JJ, Raito K (2019) Evidence of the impacts of metal mining and the effectiveness of mining mitigation measures on social–ecological systems in Arctic and boreal regions: a systematic map protocol. Environ Evid 8(1):9

Hansen RN (2015) Contaminant leaching from gold mining tailings dams in the Witwatersrand basin, South Africa: a new geochemical modelling approach. Appl Geochem 61:217–223

Ilyas S, Kim H, Srivastava RR (2021) Sustainable Urban mining of precious metals.1st edn. CRC Press

Ilyas S, Kim MS, Lee JC (2018) Integration of microbial and chemical processing for a sustainable metallurgy. J Chem Technol 93(2):320–332

Ilyas S, Lee JC (2018) Gold metallurgy and the environment. CRC Press, Taylor & Francis

Kammen DM, Hassenzahl DM (2001) Should we risk it?: exploring environmental, health, and technological problem solving. Princeton University Press, Princeton

Kamunda C, Mathuthu M, Madhuku M (2016) Health risk assessment of heavy metals in soils from Witwatersrand Gold Mining Basin, South Africa. Int J Environ Health Res 13(7):663

Lam EJ, Cánovas M, Gálvez ME, Montofré ÍL, Keith BF, Faz Á (2017) Evaluation of the phytoremediation potential of native plants growing on a copper mine tailing in northern Chile. J Geochem Explor 182:210–217

Li T, Liu Y, Lin S, Liu Y, Xie Y (2019) Soil pollution management in China: a brief introduction. Sustainability 11(3):556

Malm O (1998) Gold mining as a source of mercury exposure in the Brazilian amazon. Environ Res 77(2):73–78

Marcantonio RA, Field SP, Sesay PB, Lamberti GA (2021) Identifying human health risks from precious metal mining in Sierra Leone. Reg Environ Chang 21:1–12

Marsden DD (1986) The current limited impact of Witwatersrand gold-mine residues on water pollution in the Vaal River system. J South Afr Inst Min Metall 86(12):481–504

MINAM (2010) Guía de Evaluación de Riesgos Ambientales. MINAM, Magdalena del Mar, Peru

More K, Wolkersdorfer C (2019) Disruptive technologies in mine water management–the future. In: Wolkersdorfer C, Khayrulina E, Polyakova S, Bogush A (eds) Mine water–technological and ecological challenges. Perm, pp 597–602

More KS, Wolkersdorfer C, Kang N, Elmaghraby AS (2020) Automated measurement systems in mine water management and mine workings—a review of potential methods. Water Resour Ind 24:100136

Mudd GM (2007) Global trends in gold mining. Resour Policy 32(1–2):42–56

Ngole-Jeme VM, Fantke P (2017) Ecological and human health risks associated with abandoned gold mine tailings contaminated soil. PloS One 12(2):e0172517

Naicker K, Cukrowska E, McCarthy TS (2003) Acid mine drainage arising from gold mining activity in Johannesburg, South Africa. Environs Environ Pollut 122(1):29–40

Ouellet SM, Dettmer J, Olivier G, DeWit T, Lato M (2022) Advanced monitoring of tailings dam performance using seismic noise and stress models. Commun Earth Environ 3(1):301

Paniagua-López M, Vela-Cano M, Correa-Galeote D, Martín-Peinado F, Garzón FM, Pozo C, González-López J, Aragón MS (2021) Soil remediation approach and bacterial community structure in a long-term contaminated soil by a mining spill (Aznalcóllar, Spain). Sci Total Environ 777:145128

Popovic V, Miljkovic J, Subic J, Jean-Vasile A, Adrian N, Nicolăescu E (2015) Sustainable land management in mining areas in Serbia and Romania. Sustainability 7:11857–11877
Porcella DB, Ramel C, Jernelov A (1997) Global mercury pollution and the role of gold mining. Water Air Soil Pollut 97(3–4):205–207
RAIS (2021) Toxicity profiles. Risk Assessment Information System
Rapant S, Dietzová Z, Cicmanová S (2006) Environmental and health risk assessment in abandoned mining area, Zlata Idka, Slovakia. Environ Geol 51:387–397
Rivera-Parra JL, Beate B, Diaz X, Ochoa MB (2021) Artisanal and small gold mining and petroleum production as potential sources of heavy metal contamination in Ecuador: a call to action. Int J Environ Res Public Health 18:2794
Rodríguez-Eugenio N, McLaughlin M, Pennock D (2018) Soil pollution: a hidden reality. FAO
SA-DEA (2010) The framework for the management of contaminated land, South Africa. Republic of South Africa Department of Environmental Affairs, Cape Town
Salgado-Almeida B, Falquez-Torres DA, Romero-Crespo PL, Valverde-Armas PE, Guzmán Martínez F, Jiménez-Oyola S (2022) Risk assessment of mining environmental liabilities for their categorization and prioritization in gold-mining areas of Ecuador. Sustainability 14:6089
SENAGUA (2011) Informe Técnico Muestreo y Análisis de La Calidad Del Agua En La Cuenca Del Río Puyango. SENAGUA, Quito, Ecuador
Singh R, Singh S, Parihar P, Singh VP, Prasad SM (2015) Arsenic contamination, consequences and remediation techniques: a review. Ecotoxicol Environ Saf 112:247–270
Srivastava RR, Rajak DK, Ilyas S, Kim H, Pathak P (2023) Challenges, regulations, and case studies on sustainable management of industrial waste. Fortschr Mineral 13(1):51
Tóth G, Hermann T, Da Silva MR, Montanarella L (2016) Heavy metals in agricultural soils of the European Union with implications for food safety. Environ Int 88:299–309
UK-DoE (2009) Contaminated land exposure assessment: heavy metal guidelines in soil, UK. UK Department of Environment
US-EPA (2011) Default use of body weight 3/4 as the default method in derivation of the oral reference dose. US Environmental Protection Agency, Washington, DC
US-EPA (2014) Framework for human health risk assessment to inform decision making. US Environmental Protection Agency, Washington, DC
US-EPA (2018) Regional screening levels (RSLs)—resident soil tables [data and tools]. US Environmental Protection Agency
US-EPA (2019) EPA facts about thorium. US Environmental Protection Agency, Washington, DC
USEPA (1998) Ecological risk assessment guidance for superfund: process for designing and conducting risk assessments. USEPA, Washington, DC
Vela-García N, Guamán-Burneo MC, González-Romero NP (2019) Efficient bioremediation from metallurgical effluents through the use of microalgae isolated from the Amazonic and highlands of Ecuador. Rev Int Contam Ambient 35:917–929

Assessment and Mitigation of Environmental Footprints for Energy-Critical Metals Used in Permanent Magnets

Humma Akram Cheema, Muhammad Farhan, and Hyunjung Kim

Abstract Critical raw materials (CRMs) include cobalt (Co) and rare earth elements (REEs) that serve as essential elements in many modern, rapidly evolving clean energy technologies, from wind turbines and electrical networks to electric automobiles. As clean energy transitions accelerate, consumption of these minerals will rise significantly. The rapid demand for CRMs in technology raises serious concerns regarding supply availability and consistency. Rare earth elements (REEs) are the key elements of permanent magnets that are employed in wind turbines and e-vehicles. The demand for these CRMs is expected to increase as the use of permanent magnets in various applications continues to grow. Hence, there is a need for sustainable and responsible sourcing of these materials to ensure their availability in the long run. This chapter highlights the energy-critical metals used in permanent magnets, particularly focusing on neodymium-iron-boron (NdFeB) magnets for clean energy production. Moreover, the assessment of their economic importance and environmental challenges is also discussed in detail. Finally, mitigation approaches to environmental footprints such as recycling and reusing secondary resources, sustainable processing, and substitution techniques during the manufacturing of magnets are also discussed.

Keywords Critical raw materials · Permanent magnets · Neodymium · Cobalt

1 Introduction

Critical raw materials (CRMs) are those materials that are considered essential to the country's economy but are also vulnerable to supply disruptions. These materials are essential to many industrial processes and products, ranging from high-tech electronics to renewable energy technologies (Månberger 2023). The criticality of

H. A. Cheema · M. Farhan · H. Kim (✉)
Department of Earth Resources and Environmental Engineering, Hanyang University, Seoul, Republic of Korea
e-mail: kshjkim@hanyang.ac.kr

P. Pathak et al. (eds.), *Anthropogenic Environmental Hazards*,
https://doi.org/10.1007/978-3-031-41013-0_2

elements is not defined by their natural abundance; rather, they are classified because:

- They have a substantial economic role in industries including consumer electronics, environmental technology, automobiles, aircraft, defense, and health care.
- They face a high supply risk as a result of their high reliance on imports and the concentration of some critical raw resources in specific regions.
- Due to these materials' highly distinctive and dependable qualities for current and prospective uses, there are few feasible replacements.

Most CRMs have distinctive qualities that make it extremely challenging to replace them in the chosen application. Therefore, if the supply of critical material is disrupted, it can lead to production delays, increased costs, and a loss of competitiveness for industries that rely on that material (IEA 2020). The European Union (EU) and the US Department of Energy have established a list of critical raw materials that are deemed essential for industrial competitiveness and sustainability, as well as for the transition to a low-carbon and digital economy (EU Commission 2017; US-DoE 2023). The EU's list includes 30 raw materials, and it is regularly reviewed and updated based on the latest market developments and supply chain risks. The criticality assessed for CRMs by the US Department of Energy is depicted in Figure 1 reveals three broad categories of criticality (i.e., critical, near critical, and not critical). Materials in the upper quadrant of the matrix—with scores of 3 or higher on both axes—are characterized as high in criticality that includes dysprosium, cobalt, gallium, graphite, iridium, and neodymium.

Asia holds significant potential in terms of critical material prospects due to its abundant reserves and growing demand. Many technologically advanced countries

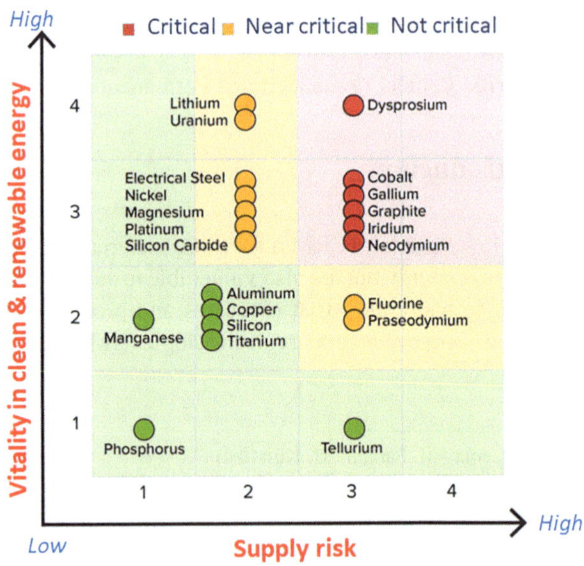

Fig. 1 Criticality matrix up to 2025, as defined by the US Department of Energy in 2023 (US-DoE 2023)

in Asia, including Japan, South Korea, and others, have a significant dependency on imports for CRMs. This is primarily because a substantial portion of global REEs production is concentrated in China, which has led to concerns about supply security and geopolitical implications. They recognize the strategic importance of energy-critical metals for their energy transition and are taking various measures to ensure a secure and sustainable supply. This includes diversifying sources, investing in domestic production and exploration, promoting recycling and circular economy practices, and advancing research and development efforts to reduce dependency and environmental impacts. This chapter particularly focuses on permanent magnet raw materials that are of significant importance due to their unique properties and wide range of applications in clean energy transition technology.

Permanent magnets are widely used in various applications such as electric vehicles, wind turbines, and electronic devices. They are made up of various materials, but the most common types of permanent magnets are neodymium based (NdFeB) and samarium-cobalt (SmCo) magnets. Both NdFeB and SmCo magnets contain critical raw materials that are essential for their performance. These critical raw materials include neodymium (Nd) that is used in NdFeB magnets. It is a critical raw material because it is in limited supply, and its extraction and processing are associated with environmental risks and concern. Dy is another rare earth element that is used in NdFeB magnets. It is a critical raw material because it is essential for improving the high-temperature performance of the NdFeB magnet. Sm is an REE that is used in SmCo magnets. It is a critical raw material because it helps in maintaining the magnetic properties of SmCo magnets at high temperatures. Co is a transition metal that is used in both NdFeB and SmCo magnets. It is a critical raw material because its supply is limited and its extraction and processing are associated with environmental concerns.

Various kinds of permanent magnets are employed in a variety of applications, most of which are crucial for developing a low-carbon economy. For example, permanent magnet synchronous generators (PMSG) are utilized in many kinds of wind turbines. A significant percentage of permanent magnets are found in the direct-drive low-speed turbine structure, which has advantages for large-scale wind turbines producing more than 5 megawatts (MW) and offshore conditions. Moreover, a large number of electric and hybrid vehicles also use permanent magnet synchronous motor technology because of their high-power density and capacity to generate high torque (Pavel et al. 2016). Ferrite, aluminum-nickel-cobalt (AlNiCo), samarium-cobalt (SmCo), and neodymium-iron-boron (NdFeB) are the four most significant permanent magnet types. NdFeB magnets are the subject of this study because they are essential for many high-tech applications and account for a sizable portion of the market for critical raw material-containing magnets (Gutfleisch et al. 2011).

The strongest permanent magnets currently available on the market are made of neodymium, iron, and boron. According to Pavel et al. (2016), the strong magneto crystalline anisotropy and high coercivity of these REE magnets are the results of the coupling of 3d transition metal with the 4f electron configuration of REEs. As a result, NdFeB magnets are utilized in a variety of high-tech applications, such as

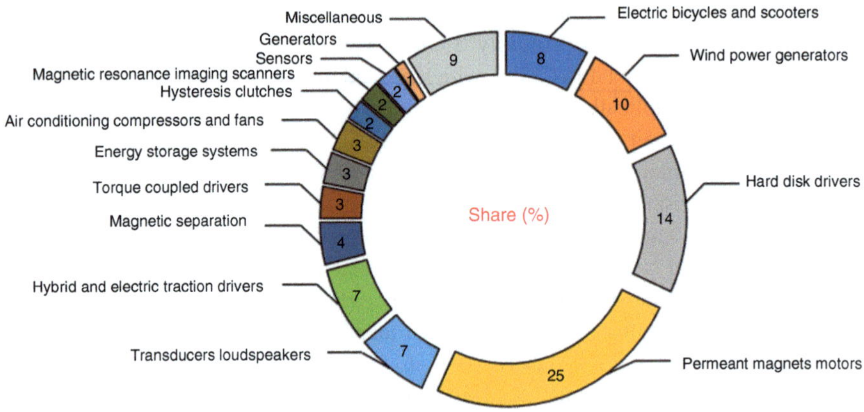

Fig. 2 The main end-use applications for NdFeB magnets

electric bicycles and consumer electronics, as well as high-performance applications like wind turbines and electric cars, where compact size and high efficiency are required (Binnemans et al. 2013). Figure 2 illustrates the main end-use applications for NdFeB magnets. They have a limited operating temperature range and are susceptible to corrosion. The composition-dependent maximum operating temperature ranges from 80 to 200 °C (Binnemans et al. 2018).

2 CRMs in Applications

CRMs are employed in many commercial permanent magnets. Table 1 lists the most significant CRM used in magnet applications as well as the percentage of magnets that use each critical raw material. The most widely used kind of permanent magnet utilized in high-tech applications is neodymium-iron-boron (NdFeB) magnets, specifically $Nd_2Fe_{14}B$. These magnets comprise mainly of iron (64.2–68.5%), neodymium (29–30%), and boron (1–1.2%), where iron can be partially replaced with aluminum, copper, and cobalt, while neodymium can be partially replaced with dysprosium, terbium, and gadolinium (EU Commission 2017; Yang et al. 2017a, b).

The specifications of the end-use application determine the exact NdFeB magnet composition. Nd and Py make up the majority of the REEs in NdFeB magnets, which typically include 31–32% of them by weight (Yang et al. 2017a, b). Py content differs and can reach a 4:1 ratio with Nd (Pavel et al. 2016). The Curie temperature (where a material loses its permanent magnetic properties) of the NdFeB magnet is raised by the addition of dysprosium and terbium, increasing the operating temperature up to 200 °C. Depending on the application, the required amount of Dy and Tb could vary from 3 to 7% in wind turbines and 9% in electric vehicles (Tercero et al. 2018). Nevertheless, these additives raise the cost of the magnet and reduce the energy density and remanence (Yang et al. 2017a, b). Thus, only applications

Table 1 Summary of CRMs used in different types of magnets and applications of NdFeB magnet

CRM	Nd	Pr	Gd	Dy	Tb	Sm	Co
Magnet type	NdFeB	NdFeB	NdFeB	NdFeB	NdFeB	SmCo	SmCo, NdFeB, AlNiCo
Wt.% share of magnets in CRM end uses	37	24	35	100	32	97	5
Applications	**Wind turbines**		**Electric vehicles**		**Electric bikes**		
	Direct drive	Geared	With PMSG		–		
Nd-Fe-B magnet	650 kg/MW	80–160 kg/WM	1–2 kg/EV		0.3–0.35 kg/bike		

requiring high-temperature stability can make use of dysprosium and terbium in higher concentrations. Gadolinium in small amounts also enhances high-temperature performance by improving the temperature coefficient (Yang et al. 2017a, b). Other rare earth elements, such as lanthanum, cerium, niobium, and gallium, have also been investigated as partial replacements for neodymium or iron to change the characteristics and improve the performance of the neodymium-iron-boron magnet (Yang et al. 2017a, b, 2012). Typical quantities of NdFeB magnets needed in different applications are compared in Table 1 (Pavel et al. 2016).

Samarium-cobalt (SmCo) magnets are primarily samarium and cobalt alloys, referred to as $SmCo_5$ or Sm_2Co_{17}. Cobalt can be partially replaced in the Sm_2Co_{17} composition by iron, zirconium, and copper. SmCo magnets have good corrosion resistance and high-temperature performance, even up to 400 °C. Samarium-cobalt magnets can only be used in a small number of applications due to the expensive cost of samarium and the labor-intensive manufacturing method (Binnemans et al. 2013). Samarium is primarily used in magnets, with 97% of that utilization taking place in the EU (EU Commission 2017). Aluminum-nickel-cobalt magnets are based on alloys that are primarily made of iron, nickel, cobalt, and aluminum. However, compared to NdFeB and SmCo magnets, these magnets have a low coercivity and a maximum operating temperature of up to 500 °C. As a result, they are only used in low-value applications and have a very small market share (Pavel et al. 2016; Gutfleisch et al. 2011). Due to its inexpensive price as well as a simple manufacturing method, ferrite magnets account for a large portion of permanent magnet sales in terms of volume (Kumar 2017; Gutfleisch et al. 2011). The most typical compositions of these magnets are $SrFe_{12}O_{19}$ and $BaFe_{12}O_{19}$, which are mostly made of Fe_2O_3 but rarely also contain Sr, Ba, or both (Riba et al. 2016). Various commercially available ferrite magnet types comprise Ln or Co to achieve enhanced magnetic characteristics.

3 Economic Significance

The economic significance of CRMs demonstrates how their consumption in applications generates economic value, according to the EC's 2017 assessment of CRMs. Magnets play a significant role in the application of many essential raw materials, as shown in Fig. 3. Magnets contribute more than half of the commercial value of neodymium, gadolinium, dysprosium, and samarium, whereas it is only slightly less than 50% for praseodymium and terbium. In the case of cobalt, even though it is utilized in SmCo magnets, which have a substantial economic impact on samarium, magnets are not one of the principal applications. Magnets are employed in many industries, including energy, transportation, and electronics, making them essential to use.

Wind turbines with permanently activated synchronous generators and direct drives, which are typically utilized in offshore installations, use neodymium-iron-boron (NdFeB) magnets. In 2019, the EU saw the construction of 3627 new offshore installations or 27.5% of total installations. Offshore installations make up 22 gigawatts (GW), or 11.5%, of the installed wind power capacity in terms of cumulative capacity (Wind Europe 2019b). Expanding offshore installations is a trend that can also affect permanent magnet demand. A total of 99.1% of offshore wind projects in the EU have been supplied by five manufacturers: Siemens Gamesa Renewable Energy (68.1%), MHI Vestas (23.5%), Senvion (4.4%), Bard Engineering (1.6%), and GE Renewable Energy (1.5%) (Wind Europe 2019a). In terms of the annual and global market share of the top offshore wind energy firms, Siemens led with 60.5%, and Vestas was second with 13.4% in 2018 (Statista 2020b). One of the most

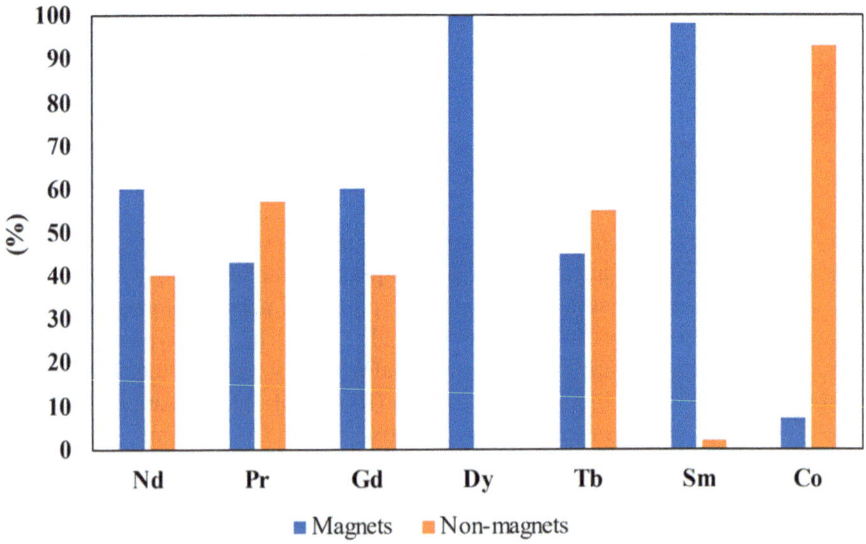

Fig. 3 Economic importance of CRMs used in magnets

important steps in reducing climate change is the electrification of the transportation industry. Permanent magnets are needed in electric vehicles' synchronous motors in addition to batteries. E-vehicles (plug-in hybrids and battery electric vehicles) now account for almost 3% of the market in Europe, up from less than 1% in 2015 (Statista 2020a). The share will keep rising in subsequent decades. The Renault Zoe, Mitsubishi Outlander Plug-In Hybrid Electric Vehicle (PHEV), and Tesla Model 3 were the top-selling electric vehicle models in Europe in 2019 with sales of 45,130, 34,600, and 95,170, respectively (Statista 2020a).

Wind energy is crucial to achieving energy and climate goals, and it is anticipated that wind power generation will continue to grow in the coming years. In 2023, the total installed capacity might reach 277 GW, according to Wind Europe's projections. This is based on the installation of an additional 88 GW of net capacity. Onshore installations are anticipated to make up at least 75% of the new installations (Wind Europe 2019a, b). EU Commission (2017) predicted that the demand for Dy, Nd, and Py by 2030 will be approximately 3.25 times greater than it was in 2015 in a minimal growth scenario and approximately 44 times higher in a rapidly growing scenario. Dy, Nd, and Py each received 71, 356, and 119 tons of demand in 2015. The most effective approaches to increasing the supply of these CRMs are substitution and recycling (Blagoeva et al. 2016). The demand for permanent magnets and electric motors will rise as more electric cars are added to the fleet, lowering emissions from transportation. According to Statista, 16.5 million e-vehicles will be manufactured by 2030 (Statista 2020b). It is predicted that battery electric vehicles (BEVs) and plug-in hybrid electric vehicles (PHEVs) will account for 3.9% to 13.0% and 6.7 to 22.1% of new automobile registrations in 2030, respectively. Both battery electric and plug-in hybrid e-vehicle new registrations have increased significantly, even though they accounted for only 1.5% of the market in 2017 (EEA 2018). According to the EC's projections for the consumption of magnets used in electric traction motors, the demand in 2030 for Dy and Py would be approximately 9.4 times higher compared with the 27 tons used in 2015 and neodymium 9.5 times higher than the 80 tons employed in the same year (EU Commission 2017). Regarding wind turbines, Blagoeva et al. found that substitution and recycling were the best ways to increase the supply of Dy, Nd, and Py for traction motors (Blagoeva et al. 2016).

4 Environmental Challenges

Critical materials have significant environmental impacts throughout their lifecycle, from mining and extraction to manufacturing and disposal. Many critical materials are extracted through mining processes, which can lead to environmental damage and degradation. Mining can cause deforestation, soil erosion, and water pollution and disrupt ecosystems and habitats for wildlife. Additionally, the mining of critical materials can release toxic chemicals and heavy metals into the environment, posing a risk to human health and the environment. The processing and manufacturing of critical materials into products can also have environmental impressions. For example, the production of electronics, which often contain critical materials such as REEs, can result in significant greenhouse gas emissions and other toxic pollutants. The use of energy-intensive processes and fossil fuels can contribute to climate change and air and water pollution. Finally, the disposal of critical material products at the end of their life can have environmental imitations. Many critical materials are not biodegradable and can persist in the environment for centuries, contributing to waste and pollution. Improper disposals, such as dumping in landfills or incineration, can release toxic chemicals and heavy metals into the air and soil, posing a risk to human health and the environment.

The virgin CRMs used in magnets have shown major risks associated with ecology (Wulf et al. 2017). The potential hazards of rare earths, boron, and cobalt as the major critical elements in magnets have been listed as two to three indicators. They have an elevated chance of contamination at the mining site, REEs have a high risk of natural disasters, and mining for boron has a high danger of stressing the water supply. Most of them are extracted from open-cast mines, which have an immediate negative effect on the environment because topsoil and vegetation are removed. If adequate measures aren't taken to mitigate these hazards, environmental effects are very likely to take place. Permanent magnets made of neodymium-iron-boron and samarium-cobalt are mostly composed of REEs such as Dy, Nd, Py, and Sm. Since the manufacture of REEs is material- and energy-demanding and produces huge quantities of emissions into the air and water as well as solid waste, it significantly harms the environment (Vahidi et al. 2016). Concerns regarding heavy metal and radioactive exposure to rivers, groundwater, soil, and the air near mine sites have been raised by their production in China, which is the largest producer worldwide (Adnan et al. 2022). Furthermore, the roasting of ores has an effect because it requires a significant amount of heat, which is provided by coal in China (Zaimes et al. 2015).

The Democratic Republic of the Congo (DRC), where more than half of the world's cobalt is mined, is reportedly problematic from both an environmental and socioeconomic standpoint (Chen et al. 2020). For instance, child labor is reportedly frequent in small-scale and artisanal mining in the DRC (O'Driscoll 2017), and mining communities' exposure to Co can cause health issues (Dunn et al. 2015). According to Dunn et al. (2015), there have been reports of heavy metal pollution of the soil, wetland acid rain, loss of biodiversity, flora dieback, and soil erosion. Arsenide ores are also extracted in cobalt-dominant mines, which might cause significant problems for the environment and public health (Dunn et al. 2015). China is home to the majority of the world's cobalt refineries (Zhang et al. 2021). Some of the possible environmental issues associated with Co refining include the processes' high-energy requirements (Farjana et al. 2019), which are made worse by China's largely coal-based electricity production, and the CO2 emissions caused by thermal decomposition during the final (calcination) step of the production of battery-grade cobalt oxide powder (Kelly et al. 2020). Boron and arsenic contamination of surface and groundwater has been caused by improper treatment of waste from borate mining activities in Turkey (Omwene et al. 2019; Gemici et al. 2008).

Due to the mining of colemanite, which contains arsenic minerals, arsenic pollution only exists in the Emet and Orhaneli stream basins (Omwene et al. 2019).

REEs, which make up 50–100% of the total impact depending on the impact category, are the primary drivers behind the environmental effects of the manufacturing process of NdFeB magnets according to Jin et al. (2018). Additionally, these researchers discovered that the effect of REEs varies significantly depending on the location of production. The environmental effects of rare earth oxides from Bayan Obo, China, were the most severe, compared to significantly lower effects from Mount Weld, Australia, particularly from Mountain Pass, and California, USA, which shut down in 2015 (Schreiber et al. 2019). Other critical raw minerals including cobalt, boron, and gallium have little to no environmental impact (Jin et al. 2018). As a result, the essential raw materials, particularly REEs, have a major impact on how magnets influence the environment. However, compared to the overall environmental effect of the end-use application, the negative effects of the magnets themselves may not be very significant. According to Nordelöf et al. (2019), who studied the ecological consequences of an electric vehicle traction motor, the role of NdFeB and SmCo magnets was minimal, aside from resource depletion, and the overall ecological impacts were mainly triggered by the manufacturing of the motor's other components and its use. Only the electricity is required to compensate for losses imposed by the motor during use—for instance, when carrying its own mass.

Figure 4 illustrates how the critical raw materials in a magnet contribute to the climate change impact of an electric car motor. The magnets' qualities also have an impact on how many other materials are utilized in the construction of the motor, which has a secondary effect on the environmental impact (Nordelöf et al. 2019).

5 Mitigation Approaches

Permanent magnets rely on critical materials that are limited in supply, such as neodymium and dysprosium. The extraction and demand for these materials can lead to resource scarcity and geopolitical concerns. Conservation efforts, such as

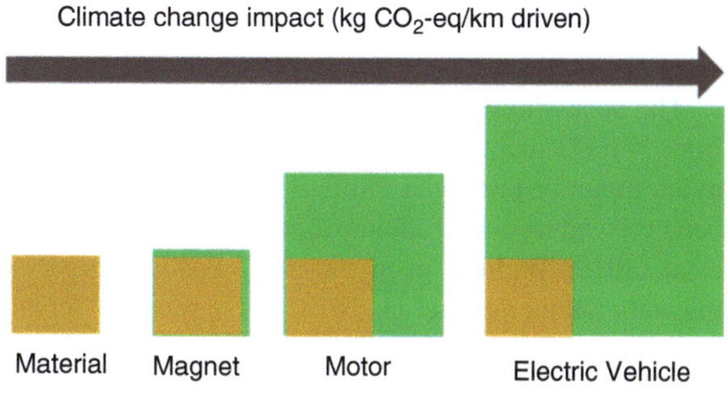

Fig. 4 Role of CRMs in a magnet and its effect on climate change of end application in an e-vehicle (Nordelöf et al. 2019; Jin et al. 2018)

recycling and efficient use of critical materials, are crucial to mitigate the environmental impacts associated with their extraction and ensure long-term availability. Mitigating the environmental footprint of permanent magnet critical materials, such as neodymium-based magnets, requires specific approaches tailored to their unique characteristics. Here are some key mitigation approaches for reducing the environmental impact of permanent magnet critical materials:

5.1 Recycling and Reusing Secondary Resources

Recycling and reusing of permanent magnets' critical materials, such as neodymium, dysprosium, and samarium, are essential for minimizing environmental impacts and conserving these valuable resources (Saito et al. 2006). To promote the circular economy for permanent magnets, efforts are being made to develop efficient collection and recovery systems. This involves establishing channels for collecting end-of-life products containing magnets, such as hard drives, electric motors, and speakers, and ensuring proper dismantling and separation. Once collected, the recycling process involves separating and sorting magnets from other materials. Various techniques, such as magnetic separation, eddy current separation, and density-based separation, are employed to extract and recover the magnet materials. Recovered magnets that meet quality and performance requirements can be refurbished and reintroduced into new applications, thereby extending their lifespan and reducing the demand for new magnet production. If the recovered magnets cannot be directly reused, they can be refabricated into new magnet products. This involves processes like grinding, milling, and sintering to transform the recovered magnets into usable forms. Additionally, magnet-to-magnet recycling techniques aim to recover the magnetic materials from used magnets and utilize them in the production of new magnets. In cases where the magnets cannot be reused or refabricated, the focus shifts to extracting and purifying the valuable magnetic materials. Various techniques, such as hydrometallurgical and pyrometallurgical processes, are employed to separate and recover REEs and other valuable components present in the magnets. Permanent magnets often contain critical materials, such as neodymium, dysprosium, and praseodymium, which are essential but limited resources. Recycling magnets helps recover and reintroduce these critical materials into the supply chain, reducing the reliance on primary mining and promoting resource sustainability.

The accessibility of Nd and Py from secondary resources, specifically recycling from NdFeB magnets and NiMH batteries, is the primary focus. Due to the enormous quantity produced, NdFeB magnets used in hard disk drives (HDDs) have been viewed as one of the most potent sources of secondary neodymium (Omodara et al. 2019; Widmer et al. 2015). Solid-state disks (SSDs), which do not have magnets (Binnemans et al. 2013), have recently replaced hard disk drives in some cases (Data Security Inc. 2014). In contrast to the 350 million hard disk drives produced in 2017, it was predicted in 2017 that the output of solid-state disks would

Table 2 Recycling CRMs from different EoL applications (Omodara et al. 2019; Bacher et al. 2020)

CRMs	Nd/Py	Nd	Nd/Py	Nd/Py
Application	NiMH batteries	Hard disk drives	NdFeB magnets in electronic waste	NdFeB magnet scrap
Recovery technique	Leaching	Leaching	Direct melting	Molten salt extraction
Recovery yield (%)	97–98 Nd, 89 Py	95	28 Nd, 1 Py	87 for Nd 89 Py

reach 320 million in 2020, approaching that of hard disk drives (Statista 2020a). The reclamation of Nd from HDDs is tough because the collection and sorption of HDDs and the effective separation of NdFeB magnets from other materials are complex. High recovery rates can be reached on a laboratory scale, as shown in Table 2, but according to Reimer et al. (2018), there is no commercial recycling of REEs from HDDs or other NdFeB magnet applications in operation yet. The adequate amounts of NdFeB waste to feed commercial recycling are above 1000 tons per year. However, the gradually rising amount of NdFeB waste will enable the scaling up of recycling techniques now being developed (Reimer et al. 2018). The most environmentally damaging phases of the manufacture of NdFeB magnets, such as mining, beneficiation, leaching, and solvent extraction, can be mitigated by recycling them (Jin et al. 2016a, b). At least 90% of the rare earth elements may be recovered through magnet-to-magnet recycling (Jin et al. 2016a, b), which reduces the impact relative to virgin magnets by 64–96% (Jin et al. 2018).

Other than from HDDs and e-waste, little data is available about recycling NdFeB magnets. For instance, wind turbines could offer significant potential for recovering REEs. However, it should be highlighted that since wind turbines have a 25–30-year lifespan, recycling them won't be able to significantly lower the demand for primary resources in the short or medium term (Habib and Wenzel 2014).

5.2 *Eco-friendly Extraction Methodology*

Conventional methods for extracting critical materials from permanent magnets often involve environmentally intensive processes that can have negative impacts on ecosystems, air quality, and human health (Yang et al. 2017a, b; Xu et al. 2004). However, efforts are being made to develop more eco-friendly extraction methods. One such approach is known as environmentally friendly or sustainable extraction. Conventional methods often involve acid leaching, where strong acids are used to dissolve and extract critical materials from the permanent magnets (Chowdhury et al. 2021). This process can result in the generation of large volumes of acidic wastewater, which requires careful treatment and disposal to prevent environmental contamination. They often utilize chemical reagents, such as solvents and reductive agents, to facilitate the separation and purification of critical materials. The use of these reagents can lead to the production of toxic waste and pose risks to both the environment and human health. Moreover, high-temperature processes, such as roasting or smelting, to separate and recover critical materials require significant energy inputs and can release harmful emissions, including greenhouse gases and air pollutants (Maroufi et al. 2017). Therefore, sustainable mining is a potential source to mitigate the environmental footprints.

Sustainable extraction methods often involve hydrometallurgical processes that use water-based solutions and mild conditions to extract critical materials. These processes have a lower environmental impact compared to high-temperature and chemical-intensive methods. Bioleaching is a promising eco-friendly extraction method that involves the use of bacteria or fungi to extract critical materials from permanent magnets. The microorganisms can selectively dissolve and recover the target metals, reducing the need for harsh chemicals and minimizing waste generation. Some other methods focus on utilizing environmentally benign solvents or green solvents, such as bio-based solvents or ionic liquids, for the extraction of critical materials. These solvents are less harmful to the environment and can be more easily recycled or reused. Eco-friendly extraction methods prioritize minimizing environmental impacts by reducing the use of harmful chemicals, energy consumption, and waste generation. These approaches emphasize sustainable practices, resource conservation, and circular economy principles. Continued research and development efforts are needed to advance eco-friendly extraction technologies and make them economically viable for large-scale implementation.

5.3 Substitutionary Techniques During Manufacturing

The quantity of CRMs in NdFeB magnets can be decreased by partially substituting REEs, by component substitution, or by using alternative magnet technologies that do not require CRMs. All of these approaches are the focus of numerous research initiatives (Tercero et al. 2018).

5.3.1 Partial Substitution of REEs in NdFeB Magnet

Instead of replacing all rare earth elements, current research proposes to minimize their content in NdFeB magnets (Omodara et al. 2019). However, it is anticipated that neodymium and praseodymium levels would significantly drop in NdFeB magnets in the upcoming 10 years (Lacal-Arántegui 2015). Several attempts to replace REEs in NdFeB magnets embrace selective co-doping of Nd with Cr and Fe with Co, which has been reported to abolish the requirement for doping the alloy with dysprosium and is believed to result in an extremely strong permanent magnet identical to the traditional neodymium-iron-boron one (Pathak et al. 2015). A similar strategy has been reported, in which lanthanum and cerium are used as substitutes for some of the neodymium (Jin et al. 2016a, b). An approach to production using a unique process that has been established by Hitachi Metals, in which dysprosium is diffused into the magnet materials rather than direct alloying, is another method for lowering the dysprosium content (Widmer et al. 2015). High-performance REE-free magnets are not now commercially accessible (Barros et al. 2019), although certain attempts have been undertaken to partially substitute Nd and Py in NdFeB magnets with other more plentiful REEs like Ce and Ln. Since these essential raw minerals are mined at the same locations, this isn't carried out for reasons of sustainability and does not completely eliminate environmental dangers. However, due to the higher concentration of Ce and Ln in ore (Haque et al. 2014), a small number of ore would require to be mined and processed in order to produce a comparable quantity of essential raw material, which could mitigate the overall environmental effect of the manufacturing of magnet raw materials. But this relies on how much Ce and Ln are required to substitute the Nd and Py, as well as any potential effects on the performance of the magnets. For instance, if the Ce and Ln magnet is heavier, the increased energy usage of an electric car during operation may cancel out any environmental benefits from the magnets' production. No life cycle analyses of the associated effects could be discovered because research on the partial replacement of REEs in NdFeB magnet designs is in its initial stages.

5.3.2 Component Substitution

Electric traction motors use NdFeB magnets, which enable the production of extremely powerful magnetic fields from minuscule volumes. The alternate method involves using electromagnets to produce magnetic fields by running electricity via a conducting coil (Widmer et al. 2015). In contrast to motors that use neodymium-iron-boron magnets, these motors are made differently (Omodara et al. 2019). Although synchronous permanent magnet rotors comprised of rare earth elements are often found in electric motors for electric vehicles, there are alternative ways for developing electric motors that do not need these rotors (Riba et al. 2016) as shown in Fig. 5. According to a study by Riba et al. (2016), some rare earth element-free electric motors can match the performance of modern rare earth electric motors in terms of, for instance, torque density and efficiency. A comparison of a few of these motor types is shown in Table 3.

Another significant application for permanent magnets that allows for component substitution is wind turbine generators. In fact, technologies reliant on permanent magnets, such as permanent magnet synchronous generators (PMSGs), developed before doubly fed induction generators (DFIGs), the most popular turbine technology using electromagnets (Pavel et al. 2016). Figure 6 illustrates the categorization of these and other principal wind turbines.

Despite the fact that DFIG generators are still the most popular choice for wind turbines overall, PMGS generators are more popular in over 5 MW generators and offshore wind turbines because they are less expensive to maintain and have higher efficiency than DFIG generators (Smith and Eggert 2018; Pavel et al. 2016). Senvion's DFIGs and Enercon's electrically excited synchronous generators

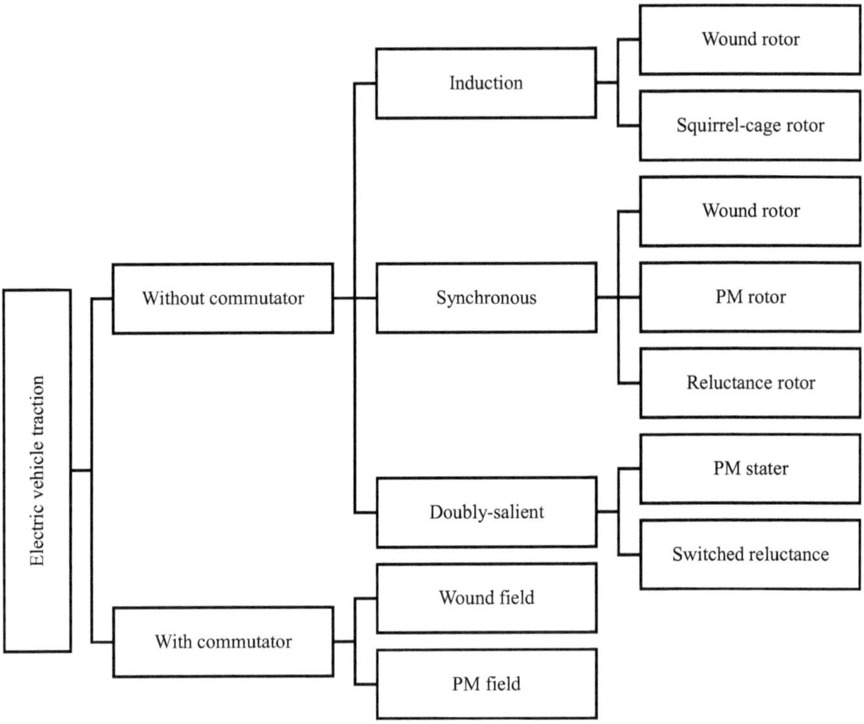

Fig. 5 Various electric motor types

Table 3 Types of motors free of REE-based magnets (Omodara et al. 2019; Widmer et al. 2015)

Motor type	Squirrel cage-type induction motor (asynchronous motor)	Synchronous wound rotor motor	Synchronous reluctance motor	Switched reluctance motor
Advantages	Ease of design, low maintenance, low cost, reliability, ruggedness	High starting torque	Ease of production	Ease of design, production and assembly, low cost, low inertia, good thermal properties, high torque density, robustness
Disadvantages	Difficulty of low-speed operation control, low efficiency, risk of overheating	Complexity of system, high cost, risk of overheating	Low efficiency, low power factor, and low torque density	Complex control, low efficiency, strong vibration, acoustic noise
Examples	Mercedes-Benz, Toyota, Tesla	Renault	ABB (industrial applications)	Rocky Mountain Technologies, US Motors (industrial applications)

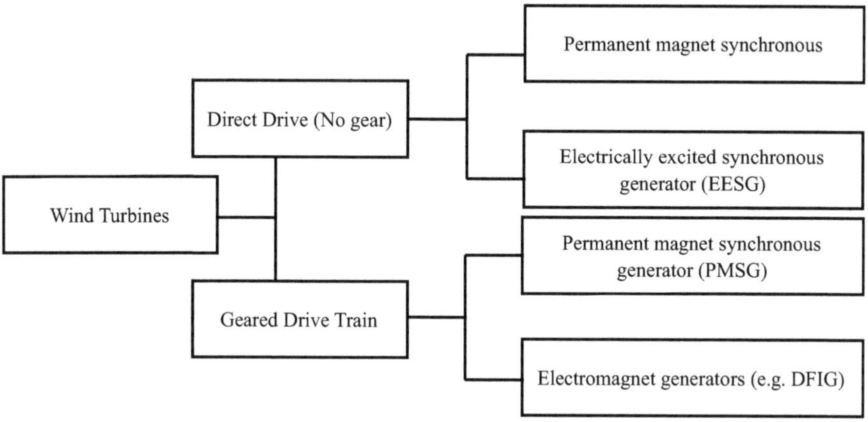

Fig. 6 Categorization of typical wind energy systems

(EESGs) are two examples of wind turbine generators used in the commercial sector without permanent magnets (Pavel et al. 2016). Component substitution may occasionally be more environmentally friendly, even while reducing environmental impacts is not the main objective. The ecological effects of e-car traction motors with three distinct core types were compared by Nordelöf et al. (2019) as:

- A neodymium-iron-boron (NdFeB) permanent magnet synchronous machine
- A synchronous permanent magnet device with samarium-cobalt magnets
- A strontium-ferrite (Srferrite) magnet-equipped permanent magnet-assisted synchronous reluctance machine (PM-assisted SynRM)

5.3.3 Magnet Technology Substitution

SmCo and AlNiCo magnets, which are likewise based on NdFeB magnets but employ different critical materials, are the contemporary alternatives to these types of magnets. Additionally, these substitutes performed inferior in terms of magnetic characteristics (Omodara et al. 2019). In some applications, ferrite magnets may be able to take a substitute for NdFeB magnets. These are primarily composed of Fe_2O_3, although they also often contain Sr, Ba, or both. According to Riba et al. (2016), ferrite magnets are considered to be the most capable alternative to NdFeB magnets in e-automobiles. According to Liu et al. (2018), one of the most promising candidates for REE-free magnets that might be employed in applications with relatively low temperatures—below 150 °C—is iron nitride, known as $Fe_{16}N_2$. Manganese-based compounds, particularly manganese alloyed with aluminum, bismuth, or gallium, have drawn interest as another class of permanent magnets (Yang et al. 2018). Magnets made on strontium-ferrite typically lack essential raw ingredients. According to Nordelöf et al. (2019), the most effective SynRM uses strontium-ferrite magnets and PM assistance, which has a reduced usage phase impact and lower carbon dioxide equivalent emissions. Strontium-ferrite magnet manufacturing has a lower climate change impact than NdFeB and SmCo magnet production, although this difference is negligible in comparison to the impact of the motor manufacturing and use phases (Nordelöf et al. 2019). Almost all impact categories investigated yielded similar findings; PM-assisted SynRM magnets execute best due to their lesser impact during use, whereas a permanent magnet synchronous machine with NdFeB magnets and a permanent magnet synchronous machine with SmCo magnets have approximately the same impact (Nordelöf et al. 2019). Although some life cycle analyses of these motors for industrial applications are available, no research on the environmental effects of alternate substitutes for NdFeB magnets in e-vehicles was found. As previously indicated, many are less effective than NdFeB magnets, which might lead to higher overall effects for the application, even if the impacts of magnet manufacture or other solutions are lower. Ozoemena et al. (2018) assessed various chances for technical advancement with respect to wind turbines. One of them is the use of permanent magnet generators to replace copper-wound rotors. Because copper has a high-energy density, the authors discovered that a permanent magnet generator turbine has a lesser impact than one with Cu-wound rotors in all impact categories.

6 Conclusions

NdFeB magnets are the most critical type of permanent magnet used in high-tech products like wind turbines and electric cars. The extraction of critical metals (mainly REEs and cobalt) can have significant environmental impacts on habitats, soil, and water consumption and depletion, air pollution, energy consumption, and emissions of greenhouse gases, along with the generation of secondary waste stockpiles in terms of mine tailings and slags. To address such issues, the adoption of various mitigation approaches, such as recycling and reuse of secondary resources, eco-friendly mining activities, and substitution or reduction of their use in magnets, is highly recommended. Effective collection systems and advanced recycling technologies are vital, along with creating awareness among industries, consumers, and stakeholders about their importance. In addition, substitution or reduction of CRMs' application in magnets can significantly control the environmental stress caused by their primary mining and metallurgical operations (in particular Nd, Dy, and Co). In this context, the application of ferrite magnets can be a better replacement for NdFeB magnets due to the lower environmental burden of iron extraction, whereas increasing the efficiency of NdFeB magnets can also be an option to reduce the environmental impact.

References

Adnan M, Xiao B, Xiao P, Zhao P, Li R, Bibi S (2022) Research progress on heavy metals pollution in the soil of smelting sites in China. Toxics 10(5):231. https://doi.org/10.3390/toxics10050231

Bacher J, Pohjalainen E, Yli-Rantala E, Boonen K, Nelen D (2020) Environmental aspects related to the use of critical raw materials in priority sectors and value chains. https://cris.vtt.fi/en/publications/environmental-aspects-related-to-the-use-of-critical-raw-material. Accessed 15 Dec 2020

Barros R, Yang Y, Bouyer E, Karhu M, Yli-Rantala E, Saarivirta EH, Samouhos M, Sundqvist L, Hu X (2019) Scrreen—D6.3. Technological gaps hindering uptake of CRMs substitution in industrial application

Binnemans K, Jones PT, Blanpain B, Van GT, Yang Y, Walton A, Buchert M (2013) Recycling of rare earth: a critical review. J Clean Prod 51:1–22. https://doi.org/10.1016/j.jclepro.2012.12.037

Binnemans K, Jones PT, Müller T, Yurramendi L (2018) Rare earths and the balance problem: how to deal with changing markets? J Sustain Metall 4(1):126–146. https://doi.org/10.1007/s40831-018-0162-8

Blagoeva DT, Alves Dias P, Marmier A, Pavel CC (2016) Assessment of potential bottlenecks along the materials supply chain for the future deployment of low-carbon energy and transport technologies in the EU. Wind power, photovoltaic and electric vehicles technologies, time frame: 2015–2030. JRC Science for Policy Report. Publications Office of the European Union, Luxembourg. https://publications.jrc.ec.europa.eu/repository/handle/JRC103778. Accessed 23 Nov 2016

Chen Z, Zhang L, Xu Z (2020) Analysis of cobalt flows in mainland China: exploring the potential opportunities for improving resource efficiency and supply security. J Clean Prod 275:122841. https://doi.org/10.1016/j.jclepro.2020.122841

Chowdhury NA, Deng S, Jin H, Prodius D, Sutherland JW, Nlebedim IC (2021) Sustainable recycling of rare-earth elements from NdFeB magnet swarf: techno-economic and environmental perspectives. ACS Sustain Chem Eng 9(47):15915–15924. https://doi.org/10.1021/acssuschemeng.1c05965

Data Security Inc (2014) SDD vs HDD: the difference between solid state storage devices and hard disk drives. Data Security Inc., Lincoln, NE, US. https://www.datasecurityinc.com/solid_state_storage_devices.html. Accessed 29 Apr 2020

Dunn JB, Gaines L, Kelly JC, James C, Gallagher KG (2015) The significance of Li-ion batteries in electric vehicle life-cycle energy and emissions and recycling's role in its reduction. Energ Environ Sci 8(1):158–168. https://doi.org/10.1039/c4ee03029j

EEA (2018) Electric vehicles from life cycle and circular economy perspectives. European Environment Agency, Copenhagen, Denmark. https://www.eea.europa.eu/publications/electric-vehicles-from-life-cycle. Accessed 2 June 2020

EU Commission (2017) Study on the review of the list of critical raw materials. Critical raw materials factsheets. Publications Office of the European Union, Luxembourg. https://doi.org/10.2873/398823

Farjana SH, Huda N, Mahmud MP (2019) Life cycle assessment of cobalt extraction process. J Sustain Min 18(3):150–161. https://doi.org/10.1016/j.jsm.2019.03.002

Gemici Ü, Tarcan G, Helvacı C, Somay AM (2008) High arsenic and boron concentrations in groundwaters related to mining activity in the Bigadi?? Borate deposits (Western Turkey). J Appl Geochem 23:2462–2476. https://doi.org/10.1016/j.apgeochem.2008.02.013

Gutfleisch O, Willard MA, Brück E, Chen CH, Sankar SG, Liu JP (2011) Magnetic materials and devices for the 21st century: stronger, lighter, and more energy efficient. Adv Mater 23(7): 821–842. https://doi.org/10.1002/adma.201002180

Habib K, Wenzel H (2014) Exploring rare earths supply constraints for the emerging clean energy technologies and the role of recycling. J Clean Prod 84(1):348–359. https://doi.org/10.1016/j.jclepro.2014.04.035

Haque N, Hughes AE, Lim K, Vernon C (2014) Rare earth elements: overview of mining, mineralogy, uses, sustainability and environmental impact. Resources 3:614–635. https://doi.org/10.3390/resources3040614

IEA (2020) The role of critical world energy outlook special report minerals in clean energy transitions. https://iea.blob.core.windows.net/assets/ffd2a83b-8c30-4e9d-980a-52b6d9a86fdc/TheRoleofCriticalMineralsinCleanEnergyTransitions.pdf. Accessed 1 May 2021

Jin H, Afiuny P, Dove S, Furlan G, Zakotnik M, Yih Y, Sutherland JW (2018) Life cycle assessment of neodymium-iron-boron magnet-to-magnet recycling for electric vehicle motors. Environ Sci Technol 52(6):3796–3802. https://doi.org/10.1021/acs.est.7b05442

Jin H, Afiuny P, McIntyre T, Yih Y, Sutherland JW (2016a) Comparative life cycle assessment of NdFeB magnets: virgin production versus magnet-to-magnet recycling. Procedia CIRP 48:45–50. https://doi.org/10.1016/j.procir.2016.03.013

Jin J, Zhang Y, Ma T, Yan M (2016b) Mechanical properties of La-Ce-substituted Nd-Fe-B magnets. IEEE Trans Magn 52(7):2100804. https://doi.org/10.1109/TMAG.2016.2524019

Kelly JC, Dai Q, Wang M (2020) Globally regional life cycle analysis of automotive lithium-ion nickel manganese cobalt batteries. Mitig Adapt Strat Glob Chang 25:371–396. https://doi.org/10.1007/s11027-019-09869-2

Kumar A (2017) Magnets and magnet materials. BCC Research, Wellesly, MA, US. https://www.bccresearch.com/market-research/advanced-materials/magnets-magnet-materials-markets-report.html. Accessed 1 Oct 2017

Lacal-Arántegui R (2015) Materials use in electricity generators in wind turbines—state-of-the-art and future specifications. J Clean Prod 87:275–283. https://doi.org/10.1016/j.jclepro.2014.09.047

Liu S, Fan HR, Yang KF, Hu FF, Rusk B, Liu X, Li XC, Yang ZF, Wang QW, Wang KY (2018) Fenitization in the giant Bayan Obo REE-Nb-Fe deposit: implication for REE mineralization. Ore Geol Rev 94:290–309. https://doi.org/10.1016/j.oregeorev.2018.02.006

Månberger A (2023) Critical raw material supply matters and the potential of the circular economy to contribute to security. Inter Econ 58(2):74–78

Maroufi S, Khayyam Nekouei R, Sahajwalla V (2017) Thermal isolation of rare earth oxides from Nd–Fe–B magnets using carbon from waste tyres. ACS Sustain Chem Eng 5(7):6201–6208. https://doi.org/10.1021/acssuschemeng.7b01133

Nordelöf A, Grunditz E, Lundmark S, Tillman AM, Alatalo M, Thiringer T (2019) Life cycle assessment of permanent magnet electric traction motors. J Transp Environ 67:263–274. https://doi.org/10.1016/j.trd.2018.11.004

O'Driscoll D (2017) Overview of child labour in the artisanal and small-scale mining sector in Asia and Africa. K4D Helpdesk Report. Institute of Development Studies, Brighton, UK. https://www.gov.uk/research-for-developmentoutputs/overview-of-child-labour-in-the-artisanal-and-small-scale-mining-sector-in-asia-and-africa. Accessed 19 Sept 2020

Omodara L, Pitkäaho S, Turpeinen EM, Saavalainen P, Oravisjärvi K, Keiski RL (2019) Recycling and substitution of light rare earth elements, cerium, lanthanum, neodymium, and praseodymium from end-of-life applications—a review. J Clean Prod 236:117573

Omwene PI, Öncel MS, Çelen M, Kobya M (2019) Influence of arsenic and boron on the water quality index in mining stressed catchments of Emet and Orhaneli streams (Turkey). Environ Monit Assess 191(4):199. https://doi.org/10.1007/s10661-019-7337-z

Pathak AK, Khan M, Gschneidner KA, McCallum RW, Zhou L, Sun K, Dennis KW, Zhou C, Pinkerton FE, Kramer MJ, Pecharsky VK (2015) Cerium: an unlikely replacement of dysprosium in high performance Nd-FeB permanent magnets. Adv Mater 27(16):2663–2667

Pavel CC, Marmier A, Alves DP, Blagoeva D, Tzimas E, Schüler D, Buchert M (2016) Substitution of critical raw materials in low-carbon technologies: lighting, wind turbines, and electric vehicles. European Commission, Oko-Institut eV, Luxembourg

Reimer MV, Schenk-Mathes HY, Hoffmann MF, Elwert T (2018) Recycling decisions in 2020, 2030, and 2040—when can substantial NdFeB extraction be expected in the EU? Metals 8(11). https://doi.org/10.3390/met8110867

Riba JR, López-Torres C, Romeral L, Garcia A (2016) Rare-earth-free propulsion motors for electric vehicles: a technology review. Renew Sustain Energy Rev 57:367–379. https://doi.org/10.1016/j.rser.2015.12.121

Saito T, Sato H, Motegi (2006) Recovery of rare earths from sludges containing rare-earth elements. J Alloys Compd 425(1–2):145–147. https://doi.org/10.1016/j.jallcom.2006.01.011

Schreiber A, Marx J, Zapp P (2019) Comparative life cycle assessment of electricity generation by different wind turbine types. J Clean Prod 233:561–572. https://doi.org/10.1016/j.jclepro.2019.06.058

Smith BJ, Eggert RG (2018) Costs, substitution, and material use: the case of rare earth magnets. Environ Sci Technol 52(6):3803–3811. https://doi.org/10.1021/acs.est.7b05495

Statista (2020a) Electric vehicles in Europe. https://www.statista.com/study/39972/electromobility-market-ineurope/. Accessed 2 June 2020

Statista (2020b) Electric vehicles worldwide. https://www.statista.com/study/11578/electric-vehicles-statistadossier/. Accessed 2 June 2020

Tercero L, Stotz H, Otmar D, García RB, Lepe GR, Bilewska K, Osadnik M, Mazur J, Ökvist LS, Eriksson J, Hu X (2018) Critical raw material substitution profiles. Scrreen—D5.1, (730227). Publications Office of the European Union, Luxembourg. http://scrreen.eu/wp-content/uploads/2018/05/SCRREEN-D5.1-CRMprofiles.pdf. Accessed 19 Sept 2018

US DoE (2023) Critical materials assessment. US Department of Energy. https://www.energy.gov/sites/default/files/2023-05/2023-critical-materials-assessment.pdf

Vahidi E, Navarro J, Zhao F (2016) An initial life cycle assessment of rare earth oxides production from ion-adsorption clays. Resour Conserv Recycl 113:1–11. https://doi.org/10.1016/j.resconrec.2016.05.006

Widmer JD, Martin R, Kimiabeig M (2015) Electric vehicle traction motors without rare earth magnets. Sustain Mater Technol 3:7–13. https://doi.org/10.1016/j.susmat.2015.02.001

Wind Europe (2019a) Offshore wind in Europe—key trends and statistics 2019. Wind Europe, Brussels, Belgium. https://windeurope.org/about-wind/statistics/offshore/european-offshore-wind-industry-key-trends-statistics2019/. Accessed 2 June 2020

Wind Europe (2019b) Wind energy in Europe in 2019. Wind Europe, Brussels, Belgium. https://windeurope.org/aboutwind/statistics/european/wind-energy-in-europe-in-2019/. Accessed 2 June 2020

Wulf C, Zapp P, Schreiber A, Marx J, Schlör H (2017) Lessons learned from a life cycle sustainability assessment of rare earth permanent magnets. J Ind Ecol 21(6):1578–1590. https://doi.org/10.1111/jiec.12575

Xu T, Li M, Zhang C (2004) Reclamation of Nd, Dy and Co oxides from NdFeB scrap. Chin Rare Earth 25:31–34. https://doi.org/10.3969/j.issn.1004-0277.2004.02.010

Yan W, Yan S, Yu D, Li K, Li H, Luo Y, Yang H (2012) Influence of gadolinium on microstructure and magnetic properties of sintered NdGdFeB magnets. J Rare Earths 30(2):133–136. https://doi.org/10.1016/S1002-0721(12)60009-X

Yang J, Yang W, Shao Z, Liang D, Zhao H, Xia Y, Yang Y (2018) Mn-based permanent magnets. Chinese Phys B 27(11). https://doi.org/10.1088/1674-1056/27/11/117503

Yang Y, Walton A, Sheridan R, Güth K, Gauß R, Gutfleisch O, Buchert M, Steenari BM, Van Gerven T, Jones PT, Binnemans K (2017a) REE recovery from end-of-life NdFeB permanent magnet scrap: a critical review. J Sustain Metall 3(1):122–149. https://doi.org/10.1007/s40831-016-0090-4

Yang Y, Walton A, Sheridan R, Güth K, Gauß R, Gutfleisch O, Buchert M, Steenari BM, Van Gerven T, Jones PT (2017b) REE recovery from end-of-life NdFeB permanent magnet scrap: a critical review. J Sustain Metall 3(1):122–149. https://doi.org/10.1007/s40831-016-0090-4

Zaimes GG, Hubler BJ, Wang S, Khanna V (2015) Environmental life cycle perspective on rare earth oxide production. ACS Sustain Chem Eng 3(2):237–244. https://doi.org/10.1021/sc500573b

Zhang T, Bai Y, Shen X, Zhai Y, Ji C, Ma X, Hong J (2021) Cradle-to-gate life cycle assessment of cobalt sulfate production derived from a nickel–copper–cobalt mine in China. Int J LCA 26(6): 1198–1210. https://doi.org/10.1007/s11367-021-01925-x

Assessing End-of-Life Room Air Conditioner Recycling Potential for Sustainable Resource Utilization in India: A Case Study for Reducing Environmental Burden

Arpita Pandey and Rudrodip Majumdar

Abstract Increasing cooling requirements for residential space owing to the rise in global temperature have led to an increased demand for room air conditioners. The pursuit for comfort in the residential and commercial spaces accompanied by expanding economic activities has increased the market size of this consumer durable globally by about three times in 2021 as compared to the 1990 levels. The high electricity consumption by air conditioners and their contribution toward greenhouse gas emission (994 Mt CO_2) have necessitated the innovation of energy-efficient technologies. The neodymium-iron-boron rare earth permanent magnet-based motor architecture in air compressors helps in reducing electricity consumption by 58%. This study projects the growth of the residential air conditioner segment in India considering the plausible near-to-medium-term scenarios and estimates the demand for the rare earth elements used in the compressor motors in India till 2030. However, the escalated demand for rare earth materials is anticipated as a major roadblock to the aspirational 'green and energy-efficient transition' since it could lead to raw material shortage. Therefore, recycling essential components from the air conditioner units is envisaged as a suitable solution for sustaining the product supply chain. Different scenarios have been considered regarding the collection of end-of-life air conditioners while analyzing the impact of material recovery and recycling. Recovery, recycling, and reuse would likely result in the efficient utilization of critical materials with limited availability worldwide. The magnet-based rare earth elements recovered from consumer durables can be used in other value chains of commercial as well as strategic importance. The suggested closed-loop recovery and recycling of magnetic materials from the end-of-life air

A. Pandey
Centre for Resource Efficiency and Governance (CREG), The Energy and Resources Institute (TERI), New Delhi, India

R. Majumdar (✉)
Energy, Environment and Climate Change Programme (EECP), National Institute of Advanced Studies (NIAS), Bengaluru, India
e-mail: rudrodip@nias.res.in

P. Pathak et al. (eds.), *Anthropogenic Environmental Hazards*,
https://doi.org/10.1007/978-3-031-41013-0_3

compressors are aimed at strengthening an alternate supply chain for fulfilling the growing demand for energy-efficient heavy consumer durables.

Keywords Air conditioners · Inverter technology · Permanent magnets · Rare earth magnets · Recycling · Sustainable utilization

1 Introduction

According to the Provisional State of the Global Climate in 2022 report brought forth by the World Meteorological Organization (WMO), the past 8 years have witnessed an intense increase in the global temperature, and this period is being touted to be on track toward becoming the 8 warmest years on record accompanied by some extreme heat wave events (WMO 2022; Fountain and Rojanasakul 2023). This temperature rise is attributable to the ever-increasing greenhouse gas concentrations as well as the accumulated heat (IPCC 2021; The Royal Society 2022). Around 44% of the world population lives in hot climatic regions (IPCC 2022; Tuholske et al. 2021). Especially during the summer seasons, the peak temperatures in the tropical regions surpass the limit up to which the human body can thermoregulate itself to remain in the normal range of 37–37.8 °C, considered as homeostasis (i.e., a stable condition for the body) (Parkin 2021). Such situations necessitate the deployment of air conditioning (AC) equipment in order to maintain comfortable thermal limits in residential living spaces. Over the past three decades, the sales of air conditioners have been continuously increasing across the world. Globally, the stock of air conditioners (AC) has increased multifold, from an estimated 575 million units in 1990 to about 2.2 billion units in 2022 (IEA 2018, 2022a; Karkour et al. 2021). In 2020, the global AC market size was valued at USD 106.60 billion, and market insights have suggested a CAGR of about 6.2% for the upcoming decade (Grand View Research 2022; Mordor Intelligence 2022). The growing economies in the Asia-Pacific region are vulnerable to the risks posed by climate change, of which increasing ambient temperature is a pivotal condition that presents a vast opportunity for the air conditioner industry. Further, a limited *product population penetration* (PPP) (number of air conditioners per 1000 households) makes the business case stronger for the manufacturers of room air conditioners (RAC). China holds about one-third of the air conditioner product stock in use (~570 million units) and dominates the market with the largest annual production (JRAIA 2019). The emerging economies, especially India and Indonesia, have shown a sharp rise in the demand for this heavy consumer durable. The air conditioner market in India has been growing rapidly and has experienced a 15-fold increase since 1990 (IEA 2018).

However, the environmental concerns emerging from the increased use of space cooling appliances have posed critical questions to researchers and policymakers. While catering to the intended purpose of providing cooling for the indoor environment, air conditioning systems contribute to 10% of global greenhouse gas emissions. In 2021, the emissions attributable to the air conditioners were estimated at 994 Mt of CO_2 (IEA 2022b). There are two possible factors contributing to total

GHG emissions from an air conditioning device. The major one is the CO_2 emissions associated with electricity consumption (fossil fuels continue to be major contributors toward electricity generation). The other one is associated with the leakage of the refrigerants that can take place during equipment assembly, repairs, as well as recycling and disposal at the end of life. The refrigerant leakage-oriented GHG emissions from the space cooling device comprise 30.7% of total emissions over the life cycle (Dong et al. 2021).

Space cooling alone accounts for 16–20% of the final electricity consumption in the buildings and residential complexes, and the consumption was estimated at about 2000 TWh in 2021 (IEA 2018, 2022a; Karkour et al. 2021). The annual energy consumption and the emission intensity of a single AC unit with a 1.5-ton cooling capacity are estimated to be around 548 kWh and 439 kg CO_2, respectively, based on the availability of operational data pertinent to the testable conditions (Dong et al. 2021). This, however, does not account for the emissions due to the refrigerant leakages. Such high-energy consumption levels necessitate the deployment of energy-efficient components with equivalent or superior performance characteristics.

Among the various parts of an air conditioner unit, the air compressor is the major component in terms of energy usage since it provides the driving mechanical force for air circulation. Air compressors typically account for about 80% of the intrinsic power consumption (Siriwardhana and Namal 2017; Jain 2022). Therefore, the use of an energy-efficient compressor architecture would substantially dictate the performance of the overall equipment. Hence, significant research and development efforts have been directed toward improving the efficiency of compressors, and several modifications have been made to the designs as well as the material specifications for the air compressor motors. The inverter technology-based compressors, which use rare earth (NdFeB (neodymium-iron-boron)) permanent magnets in their motor architectures, help in reducing electricity consumption by about 58% (Daikin 2023).

The rare earth (RE)-based permanent magnets (predominantly NdFeB magnets) have a crucial role in the technological transition toward energy-efficient consumer appliances. The demand for the NdFeB permanent magnets has been increasing for their potential applications in electric vehicles, in wind turbine generators, as well as in various configurations of servomotors used in industrial automation, primarily owing to their superior magnetic properties (Trench and Sykes 2020; Constantinides 2012). The global consumption of permanent magnets for electronics and consumer durables was estimated to be around 26.5% of the total consumption by volume (Ma and Henderson 2021; Statista 2023a, 2016a, b). A projection made in 2012 indicated that about 2.5% of the total rare earth permanent magnet requirement in 2015 would be attributable to the air conditioner compressors and the brushless DC (BLDC) motors used in the cooling fans (Constantinides 2012). Consequently, the demand growth in this industry segment is expected to increase the future demand for rare earth-based permanent magnets. The present study aims to estimate the demand for NdFeB permanent magnets for the room air conditioner segment in India considering two plausible growth scenarios for the Indian RAC market for the period up to 2030. The demand for the different rare earth materials used in the NdFeB magnets, i.e., neodymium, praseodymium, and dysprosium, is also estimated.

As highlighted in the reports published by the IEA and other international organizations, the supply shortage of critical minerals is expected in the coming decades, since their natural deposits are geographically concentrated in a few locations. For example, the deposits for the rare earth materials used in the manufacture of permanent magnets are predominantly located in the countries like China, the United States, and Myanmar, with the share in rare earth oxide-equivalent (REO) production being 60.63%, 15.52%, and 9.38%, respectively. India came at seventh position with a 1.05% share in the REO production in 2021 (Statista 2023b; USGS 2020). However, China plays much significant role when it comes to NdFeB permanent magnet manufacturing, contributing to more than 80% of the magnet value chain (Eggert et al. 2016; Gschneidner 2015; Texas Comptroller 2021). The supply chains are going to face bottlenecks and possible disruptions owing to the volatile geopolitical ambiance in some of the critical and nodal countries and regions. Therefore, it is important to provide resilience to the global network of rare earth material supply chain by augmenting an alternate supply source, primarily through the recycling of the end-of-life (EoL) product. The sensitivity analysis has been carried out in the context of material recovery, considering two predominant magnet recycling techniques and the associated recovery rates. The results show that in the case of a 100% EoL RAC collection scenario, about 12% of the annual magnet demand could be met through recycling by 2030, and this number can even grow to 25% in the coming decade. The present analysis highlights the potential for building a sustainable business network in India comprising e-waste agencies, recyclers, as well as magnet manufacturers, and OEMs (original equipment manufacturers). The suggested approach encompassing EoL recycling of heavy consumer durables and the magnet recovery could strengthen the domestic manufacturing ecosystem while reducing the import dependency substantially.

1.1 India's AC Industry Structure

Among the heavy consumer durables or the "white goods," the room air conditioner market in India has grown rapidly over the last few years. In 2021, the total annual production of air conditioners stood at 8.1 million units. Based on the end use, it can be divided into two segments: room air conditioners (RACs)/light commercial air conditioners (LCAC) and commercial and industrial air conditioners (C&I AC). In India, the market share of the RAC segment is 78%, whereas the C&IAC segment accounts for 22% of the total AC market by product volume (Oswal 2018a). It is expected that in the upcoming decade, the market share for RACs will increase to 88%. The current population penetration of air conditioners in India is in the range of 6–7%, which is below the global average of 35% and much lower as compared to the penetration levels prevailing in China (60%) and the United States (80%) (Statista 2018; Karali et al. 2020; Khosla et al. 2021). The per capita energy consumption for space cooling in India stands at 69 kWh as compared to the world average of 272 kWh (MoEFCC 2019; AIEEE 2021). It is expected that the nation's economic growth will provide substantial impetus to the air conditioner market in India toward its expansion in the coming years. Expansion of economic activities, increase in disposable income, the quest for improved standards, and availability of different customized product options are likely to fuel AC ownership in India. Furthermore, the recent boom in the start-up sector in developing economies like India would require RACs (or LCACs) in large numbers. Therefore, in order to curb high electricity consumption as well as the greenhouse gas emissions attributable to air conditioners, demand for energy-efficient appliances for smaller commercial spaces will grow with time.

Few socioeconomic factors, such as the increasing extent of urbanization, high building densities, and changing building norms, tend to promote the use of air conditioning and ventilating equipment (De Cian et al. 2019; Davis and Gertler 2015; Pavanello et al. 2021). The air conditioning capacity and performance parameters for the indoor AC units are measured in terms of the floor area and the number of occupants in the space to be cooled (MoHUA 2016; Energy Conservation Act 2001; BEE 2006). These factors coupled with the scorching hot and humid tropical weather conditions would create a major demand for air conditioners for residential and commercial spaces (De Cian et al. 2019; Davis and Gertler 2015; Pavanello et al. 2021; Yun and Steemers 2011; Pandita et al. 2020).

The diverse product categories within the AC industry are segregated based on end use (room AC and commercial AC), machine architecture (stand-alone and split), cooling capacities (tonnage and subsequently power consumption), and compressor technology (fixed-speed and inverter). This study is focused on the residential product segment (also known as room air conditioner—RAC), which encompasses light commercial air conditioners (LCAC) as well. From the technological point of view, the inverter technology-based compressors require brushless direct current (BLDC) motors or permanent magnet synchronous motors (PMSM) which use permanent magnets in their rotors. In order to ascertain the demand for NdFeB magnets and rare earth materials (Nd, Pr, Dy) for the energy-efficient RACs/LCACs in India, the estimation of the market share of inverter technology-based AC units within the total room air conditioner segment is essential.

1.2 Scope of the Study

This study considers residential air conditioners (RACs) and light commercial air conditioners (LCACs) together as an undivided segment. In order to facilitate the ease of analysis and discussion, this undivided class is represented by the acronym *RAC*. Such a consideration is motivated by the fact that the design and the composition of the rare earth permanent magnets used in commercial and industrial air conditioners (C&I AC) are distinctly different from the segment of interest because of the higher-power requirement dictated by the application. The magnet requirement for the C&I AC segment is not covered in this study since the products are often highly customized based on the end use. There is also a scarcity of reliable data regarding the magnet quantity and composition used in the different commercial and industrial cooling applications.

This study assumes technical dominance of neodymium-iron-boron (NdFeB)-based permanent magnets in the compressor motor architecture over the ferrite counterpart, as indicated by the previously reported studies (Semones 1985; Dent 2012; Gieras 2010; Ormerod 2022). With a view to assessing the potential for recovering the magnet-based rare earth materials through the recycling of end-of-life air conditioner compressors, two different predominant recycling techniques have been considered to account for different levels of material recovery efficiency. Further, different levels of availability of the *EoL* compressor motors have been accounted for by considering three different scenarios, as discussed later. The pivotal goal of this exercise is to quantify the alternate availability of magnet-based rare earth materials via the route of recycling that could support the accelerated growth in the demand for these critical minerals. It is noteworthy that this analysis does not account for the environmental and economic costs associated with the recycling of the air conditioner compressors.

2 Literature Review

2.1 Advent of Energy-Efficient Cooling

The air conditioning system consists of four main functional components, viz. condenser coil, evaporator coil, expansion valve, and compressor (Brain et al. 2011). Among these, the compressor is an essential component responsible for the air circulation. Since the electric motor used in the compressor consumes a large fraction of the total energy required by the AC machine, it is necessary to find energy-efficient architectures for the compressor motors.

The energy efficiency ratios of the electrical machine dictate the power consumption pattern of the compressor, depending on the design and material specification of the key operating units, especially the motor and the controller (Dubas et al. 2005). The fixed-speed compressors are devoid of the automatic adjusting mechanism that is used in controlling the airflow speed and the cooling capacity. Consequently, the machine operates at a constant airflow speed even when the ambient temperature is on the lower side and the cooling requirements are less. Such constant speed operations lead to higher power consumption and may lead to excess humidity in confined spaces (Chen et al. 2009).

The variable speed compressors, which were primarily used in industrial cooling systems to adjust the input power according to the required output, offer key mechanical advantages during the operations. Since the space cooling requirements change with the changing ambient temperatures, the inverter switch-enabled motor of the compressor can improve the efficiency of the AC unit and reduce power consumption by effectively controlling the drive speed and altering the airflow speed. The inverter AC unit uses a permanent magnet (PM) in its motor architecture. PM-based motors offer advantages over their induction-based counterparts in terms of higher speed, superior energy efficiency, better power flexibility, better heat transfer mechanism, ease of maintenance, and longer operational life (Mishra et al. 2014; Shi et al. 2019). The material composition of the compressor motors and the fees associated with the patented technology make the manufacture of inverter ACs costlier.

However, recent trends indicate an increasing market share for the inverter ACs. In 2016, the inverter ACs comprised only 12% of the market. Since then, the share increased to about 80% in 2020 and is expected to increase further in the coming years (Mukherjee 2022; The Hindu Business Line 2018; Kanchwala 2022). From closer scrutiny of the market trends, it is evident that almost all commercial and industrial air conditioners (C&I ACs) use compressors equipped with variable speed technology (Ohyama and Kondo 2008; Uzhegov et al. 2017). In contrast, the market share of inverter technology-based products is about 60% in the case of the RACs (Oswal 2017a, 2018b). Considering the total cost of ownership (TCO), the consumers are progressively preferring inverter AC units despite the higher price of equipment, since these AC machines are equipped with smart technology and the reduced power requirements finally lead to substantial savings in energy bills (Jain 2022; Daikin 2023). Furthermore, the global convergence toward improved technology with higher star ratings and standardized labeling by regulatory agencies has been the pivotal enablers behind the quick adoption (Energy Conservation Act 2001; BEE 2006).

This study has analyzed the currently available models and the product trends for the major players in the Indian RAC market. It became evident that some major manufacturers, such as LG, Daikin, Carrier Midea, and Whirlpool, are the proponents of inverter-compressor technology, and their respective product lines have transitioned to the variable speed compressor designs (Daikin 2023; The Hindu Business Line 2018; Kanchwala 2022). On the other hand, a few other major players, like Blue Star, Voltas, and Samsung, have around 80% of products outlined as inverter technology-enabled. In the Indian context, the BEE (Bureau of Energy Efficiency) updated its rating norms for air conditioners in 2020, favoring compressor designs with higher energy efficiency and superior performance parameters (BEE 2020a, b; PIB 2020).

2.2 *Rare Earth Permanent Magnets: Role and Requirement*

The circulation of pressurized air for heat removal requires high speed, which necessitates the use of high-power-density electric motors in air compressors. Therefore, permanent magnets are preferred by the motor manufacturers since they facilitate higher torque generation at zero degree. There are other advantages, such as better thermal reliability, lower physical weight, lower noise, compactness, and ruggedness (Kim and Lee 2005). The NdFeB RE permanent magnets offer substantial weight reduction (by about 60%) and higher torque per unit mass as compared to the ferrite-based counterparts (Shinetsu Rare Earth Magnets 2007). The high-coercivity NdFeB magnets are available for temperatures reaching up to 220 °C in the operating environment, with negligible irreversible loss (Dent 2012; IMA 2018). The need for a low-temperature coefficient has led to the development of several grades of NdFeB magnet to meet specific operational requirements (IMA 2018; Bunting E-Magnets 2023).

The studies pertinent to system-level analysis on the PMSM (permanent magnet synchronous motor) architecture, design, and material specifications define a ratio of the magnet weight to the active weight of the motor or a minimum allowable motor power-to-active mass ratio (Campbell 2008; Collocott et al. 2004; Palmer 2022) for efficient operation. In the context of the manufacture of rare earth permanent magnets, previous studies have reported a 32–35% share of rare earth materials by weight depending on the applications (Fastenau and van Loenen 1996). Considering the internal temperature of a typical air conditioning system, compression molded magnets/bonded magnets are preferred by manufacturers for the maximum operating temperature reaching up to 170 °C (Im et al. 2021; Coney et al. 2002). The addition of dysprosium (Dy) increases the coercivity and temperature resistance of the magnet.

The relative share of neodymium (Nd), praseodymium (Pr), and dysprosium (Dy) would depend on their availability, market prices, and technological choices made by the manufacturers of magnets and compressor motors. In an efficient NdFeB permanent magnet for general application, the shares of Nd, Pr, and Dy are about 29%, 1%, and 5% by weight, respectively. Available open-source data indicate that over the past decade, the demand for permanent magnets for consumer electronics was higher than that for industrial purposes (Statista 2016a, b). However, in the near-to-medium-term future, a rapidly growing demand is expected from the industrial applications and the green energy sector (especially, the large magnets used in wind turbine generators) (Alonso et al. 2020). The growth of the electric vehicle segment would also require high-quality NdFeB magnets.

IEA reports highlight that the quest for comfort in the warming world would lead to an increasing level of use of the air conditioners, and it will be favored by the overall economic development. Increased population penetration of heavy consumer durables (including ACs) would lead to substantially higher electricity consumption in the coming decade (IEA 2022a, b). However, the higher price differential associated with energy-efficient high-end air conditioners forces medium-income group buyers to settle for inefficient products. Therefore, from a national perspective, a developing economy like India needs to enforce energy efficiency standards very strongly so that electricity consumption can be reduced and the associated emission burden (attributable to the power generation from fossil-based sources, which would continue to be the mainstay in India in the foreseeable future) on the economy.

It is noteworthy that the price fluctuation of the intermediate products and components (e.g., the rare earth permanent magnets) caused by the interruptions in the supply of raw materials may hinder the envisaged aspirational transition to more efficient products. Several reports have highlighted the concerns related to the insufficient primary supply of Nd and Dy (Habib and Wenzel 2014; Barteková 2016; Gupta and Ganesan 2014; Dutt and Tyagi 2022; IEA 2021). Therefore, it is important to consider innovative alternate approaches to ensure a sustainable and smooth transition to higher-energy efficiency with minimal discontinuity in the supply chain. One such way is the recycling of end-of-life products, which can enhance the raw material availability by catering up to 50% of the demand by 2100 (Habib and Wenzel 2014). In comparison with other consumer electronics items, on average, the quantity of magnetic materials present in a typical air conditioner unit is much higher (European Commission 2017). India, being the second largest manufacturing hub (Lu 2019) for heating, ventilation, and air conditioning (HVAC) components, has the potential to recycle magnets to recover critical minerals and reuse them for the manufacture of fresh magnets for a wide array of different applications. Such a circular approach of recycling and remanufacturing can help India in creating critical infrastructure and capabilities which can directly benefit the nation in terms of strategic strength and economic prosperity. Further, recycling would lead to a lesser extent of mining activities associated with magnet-based rare earth materials. In India, rare earths are mainly extracted from beach sand minerals (Indian Bureau of Mines 2016). Since the coastal areas in the country are vulnerable to various anthropogenic activities (Mohanty et al. 2023), the reduced extent of beach sand mining would translate into larger environmental benefits.

3 Methods

The study has estimated the demand for magnet-based rare earth materials for the RAC compressors considering two different scenarios (i.e., two different CAGR values for the sector). In 2021, the annual production of RACs in India stood at 6.32 million units, and of these, about 60% were equipped with the inverter technology. The previous production data and the cumulative stock of RAC units up to 2021 were considered to estimate future RAC market size in two growth scenarios, a conservative growth scenario of 10% CAGR and an optimistic growth scenario of 14% CAGR. A progressively increasing market share of inverter technology-based compressors is simulated by considering a time-dependent product replacement trajectory for fixed-speed technology-based compressors between 2021 and 2030. It's expected that the share of inverter RACs in the total annual production in India will increase to 90% by 2030 from a market share of 60% in 2021.

The prevalent market share information for the RAC units in India based on the cooling capacities (i.e., motor power specifications) has been used to calculate the average magnet requirement per air conditioner (see Table 1). It has been considered that 1 ton of cooling capacity is equivalent to 12,000 BTU/h or 3.5 kW. From the table, it can be seen that the weighted average magnet requirement for the residential air conditioner is about 0.115 kg. Using this value as the baseline information, the annual and the total magnet requirement are estimated for the period 2021–2030.

Table 1 Magnet requirement for RACs with different cooling capacities (Campbell 2008)

Cooling capacity (tons)	Market share	Magnet requirement (kg)
0.5–1	2%	0.071
1	37%	0.089
1.5	55%	0.126
2–3	6%	0.193
Weighted average		**0.115**

Additionally, the demand for the magnet-grade rare earth metal powders (Nd, Pr, and Dy) has been estimated assuming the average composition for NdFeB magnet mentioned earlier (see Sect. 2.2).

Further, the study incorporates the possible scenarios for recycling of air conditioner units to account for the recovery potential of magnets and rare earth materials from the compressor motors of the EoL products. It provides an assessment of the availability of magnet-based critical materials from the envisaged alternate source with a view to meeting the growing demand.

The *idealistic complete collection and recycling* (highly ambitious scenario) were assumed with a 100% collection of end-of-life compressors for extraction of all critical minerals and metals. Considering the 10-year operational life of an air conditioner compressor as claimed by the prominent manufacturers (Oswal 2017b), it was assumed that the inverter RAC units which entered the market in 2015 will be available for recycling in 2025. As per the industry insights (Oswal 2017b, c, 2018a, c), the share of inverter RACs in the Indian market was about 10% in 2015, and before that, the product share was negligible. Considering the negligible market footprint of inverter technology in RAC segment prior to 2015, the product volume of the earlier years is not considered in the calculations pertinent to the potential for recycling.

Among the commercialized methods of recycling (European Commission 2017; Constantinides 2021; Baba et al. 2013), the physical separation technology with an overall rare earth element (REE) extraction yield of 95% (Yang et al. 2017) and the hydrometallurgical process based on leaching method with an overall yield of 82% were considered (Bandara et al. 2016). As mentioned earlier, the present study considers two possible demand growth trajectories (10% and 14% CAGR) for NdFeB permanent magnets and corresponding rare earth elements (Nd, Pr, Dy) for the RAC segment. Further, the potential of alternate supply via recycling has been quantified by considering two commercially viable methods. Table 2 summarizes these scenarios. Further, in the discussion section (see Sect. 5), three different EoL product collection scenarios have been discussed to demonstrate the possible variations in the availability of critical minerals when the route of recycling is considered for bolstering the supply chain.

Table 2 Growth scenarios of RAC- and RE-based magnets and different scenarios for recovery of magnet and REEs

Scenarios for projection of demand of permanent magnets and REEs and different scenarios for estimating recycling potential			
Growth in demand for RACs and rare earth-based permanent magnets		Potential alternate supply via end-of-life RAC recycling and recovery of permanent magnets and REEs	
Conservative growth scenario (CGS) (**10% CAGR**)	Optimistic growth scenario (OGS) (**14% CAGR**)	Physical separation and extraction method (recycling yield 95%)	Hydrometallurgy—leaching method (recycling yield 82%)

4 Results

4.1 Demand Estimation for Permanent Magnet and REEs

The total annual production levels of RACs in India are estimated to reach 16.61 million units and 30.94 million units by the end of this decade for the 10% and 14% CAGR scenarios, respectively. Correspondingly, the annual production levels of inverter RACs are estimated to reach 14.94 million and 27.83 million units by 2030, respectively, for the growth rates mentioned above. The yearly total production levels of RACs and those for inverter-based RACs in the conservative and optimistic scenarios are shown in Figs. 1 and 2 for the period 2021–2030.

Corresponding to the growth in the production of the inverter RACs, the requirement for permanent magnets will also increase. The estimated values for the total magnet demand for the duration between 2022 and 2030 were found to be about 9033 and 15,174 tons under the conservative growth and the optimistic growth scenarios, respectively. The total demand for the magnet-grade powder of heavy rare

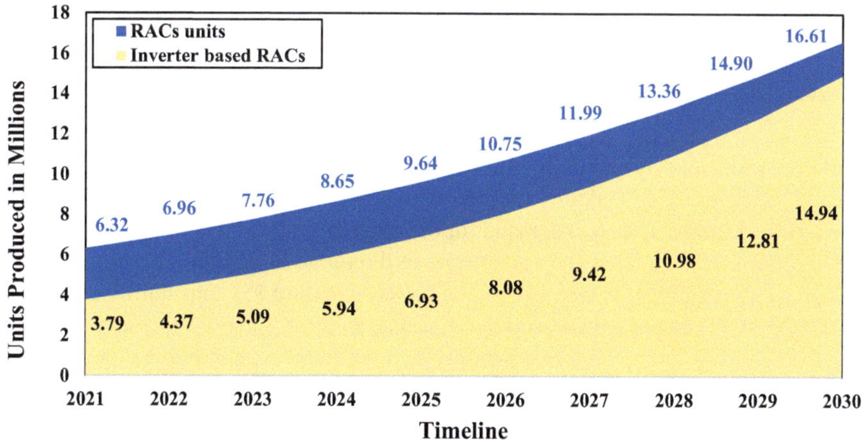

Fig. 1 Growth trajectory of RAC segment in conservative growth scenario (CGS)—CAGR 10%

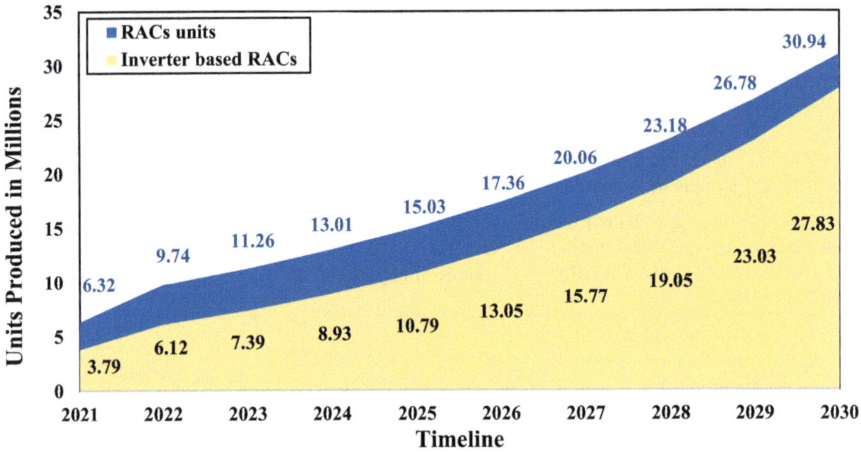

Fig. 2 Growth trajectory of RAC segment in optimistic growth scenario (OGS)—CAGR 14%

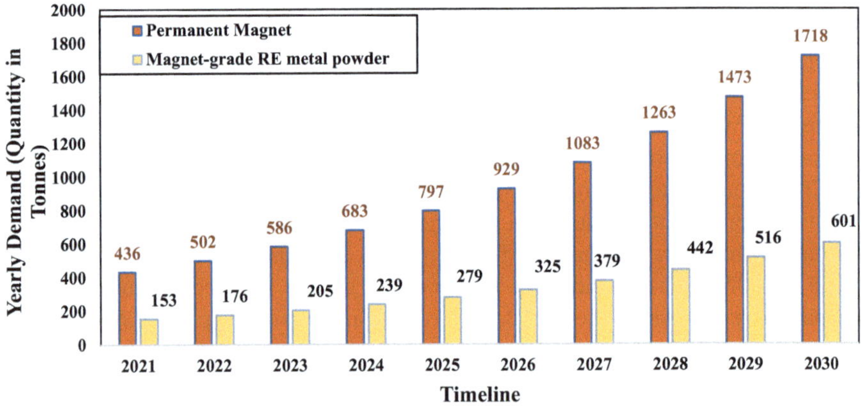

Fig. 3 Demand for permanent magnet and magnet-grade RE metal powder in conservative growth scenario (10% CAGR)

earth dysprosium (Dy) metal is expected to be in the range of 451–758 tons for the period 2022–2030, considering the two growth scenarios (10% and 14% CAGR) as the terminal limits. The total demand for the magnet-grade powder of light rare earth neodymium (Nd) metal is estimated to be in the range of 2619–4400 tons, and that for praseodymium (Pr) is estimated to be between 90 and 151 tons during the period 2022–2030, for the assumed growth scenarios. Figures 3 and 4 represent the year-wise demand for permanent magnet and the total as well as individual demand for the magnet-grade powder of three rare earth metals (Nd, Pr, and Dy) for the conservative growth scenario (CAGR of 10%). Figures 5 and 6 represent the year-wise demand for the NdFeB magnet and the magnet-grade powder of the relevant REEs (Nd, Pr, Dy) for the optimistic growth scenario (CAGR of 14%).

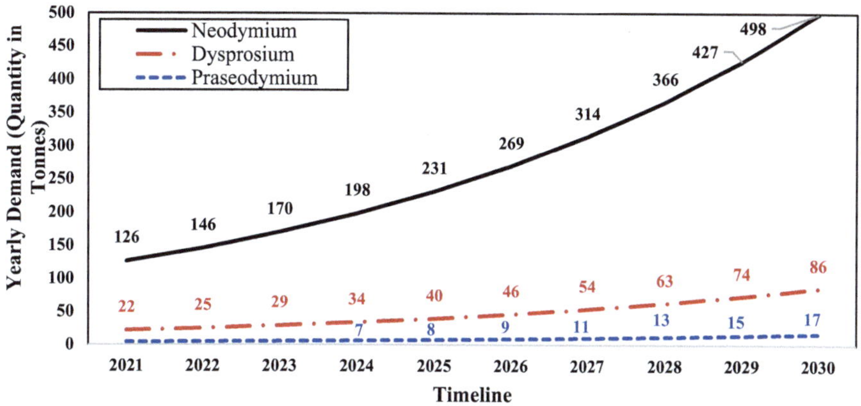

Fig. 4 Estimated demand for individual magnet-grade rare earth metal powder in conservative growth scenario (10% CAGR)

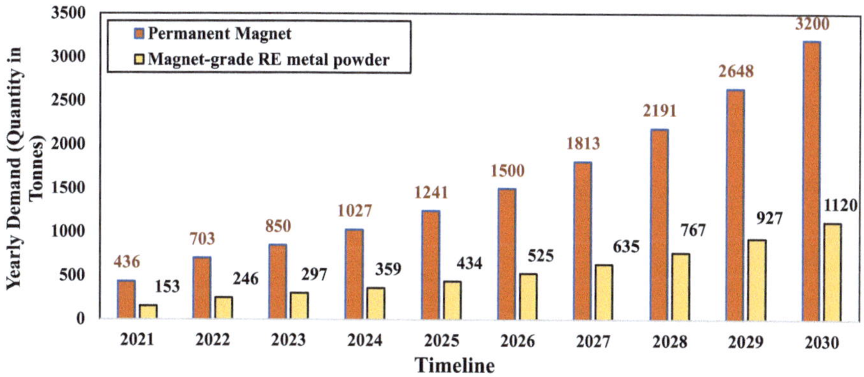

Fig. 5 Demand for permanent magnet and magnet-grade RE metal powder in optimistic growth scenario (14% CAGR)

4.2 Alternate Supply Potential for Permanent Magnet and REEs Through Recycling of EoL Products

For the assumed idealistic 100% collection scenario for end-of-life RACs (which is defined as a highly ambitious scenario (HAS)) and the adoption of two commercialized recycling methods (as mentioned in Sect. 3), a total magnet recovery amounting to about 680 tons via the physical separation process (dry extraction method) is estimated, whereas that for the hydrometallurgical process is about 586 tons, for the period 2025–2030. Figure 7 shows the year-wise quantum of magnetic material recovery during 2025–2030 for the two different recycling processes considered for the highly ambitious scenario of RAC collection.

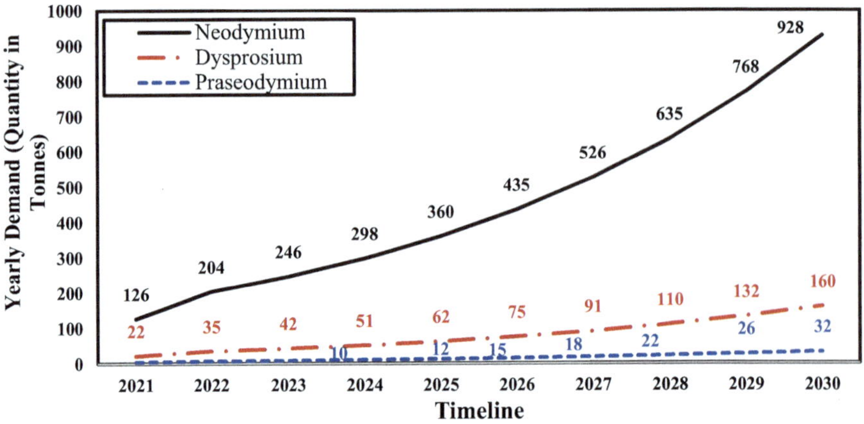

Fig. 6 Estimated demand for individual magnet-grade rare earth metal powder in optimistic growth scenario (14% CAGR)

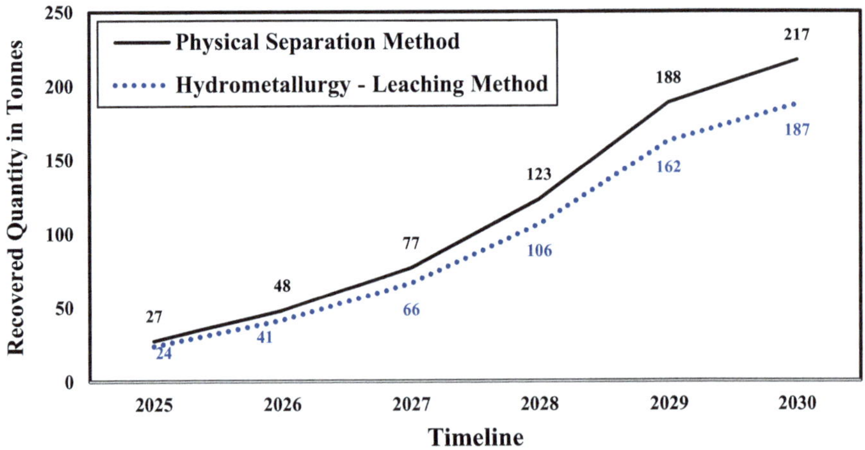

Fig. 7 Quantity of recovered permanent magnet in highly ambitious scenario (HAS) via the two predominant recycling methods

Recycling as an alternate source of raw material has limited potential in the initial years and also shows a slow pickup initially. Therefore, during the initial years of recycling (2025–2027), the contribution of recycling toward meeting the annual demand for NdFeB magnet and associated REEs would remain low (within 1–3%). However, by the end of the decade, the contribution from recycling will start becoming substantial because of the higher extent of magnet recovery, attributable to the higher market share of inverter technology-based RACs. The recycled magnet quantity may even reach up to 10–12% of annual demand levels by 2030. However, as mentioned previously, 100% collection of RACs is a highly ambitious scenario and appears to be some distance away considering the current scenario of formal electronic waste collection and recycling which comes to around 5%. Needless to mention that with a 5% collection of EoL RAC units, the recovery of permanent magnets and REEs could be negligibly small as compared to total annual demand and the recovery process may appear prohibitively expensive considering the yield.

5 Discussions

The IEA report and several studies have highlighted the emerging supply risks for the critical materials, which are considered the key drivers of the green energy transition with energy efficiency at the core. Rare earth permanent magnet (mainly NdFeB magnet) is an essential intermediate product for building high-performance energy-efficient motors for a wide array of applications of commercial importance. Since the processing of magnet-based REEs and the manufacture of RE permanent magnets have become largely monopolized in the global arena (Ma and Henderson 2021; USGS 2020; Ferreira and Critelli 2022; The Print 2021), it is imperative for the developing economies like India to explore the possibilities of developing an alternate supply source for the crucial minerals, which are used in the manufacturing of commercially important products. Recycling of end-of-life products is one such way which may prove to be a boon if the recovery and recycling-oriented industry ecosystem is planned in an integrated manner. Recycling and reuse would help in reducing the energy- and emission-intensive mining activities, which have a substantial hidden cost that is often overlooked while assessing the product's life cycle.

As highlighted before, even though the quantum of recycled permanent magnets and REEs would be low during the present decade, the impact of recycling will be much higher in the coming decades as the demand and the penetration of the energy-efficient consumer durables (including RACs) continue to grow.

Closer scrutiny of available literature pertinent to the analysis of unit-level recycling brings forth two main barriers to *organized component recycling*, (a) the technical aspects associated with the recovery of individual metals from the alloy forms extracted and (b) the scale of used products turning for recycling (collection rate of e-waste). It is observed that the recovery of individual REEs from the recycled motor is not equivalent to the quantities used during the primary manufacturing phase of the permanent magnet. The extent of REE recovery differs depending on the recycling process and demagnetization technique used, because of the physical and chemical properties of the elements (Rasheed et al. 2021). At the commercial scale, the recycling methods and processes are determined considering the resale value of the recovered RE materials in the market. For example, in the case of NdFeB magnet with dysprosium content, the heavy rare earth element dysprosium has a lesser recycling yield, and therefore it has a high value for the recyclers owing to its critical role in the high-grade permanent magnets with a higher power density (De Campos and De Castro 2020; Alves Dias et al. 2020). The extraction and recovery processes of dysprosium (Dy) are comparatively more complex and time taking. Light rare earth minerals neodymium (Nd) and praseodymium (Pr) have higher recycling yields and require relatively less intensive recycling processes (Yang et al. 2017; Bandara et al. 2016).

Considering the rising demand for Dy in RE permanent magnets used in high-performance applications, its recovery with the highest purity has been aimed for in the R&D efforts and by commercial enterprises (De campos and De Castro 2020; Fujita et al. 2022; Tunsu 2018). Reported studies highlight that dysprosium extraction has always been a challenge in NdFeB magnet recycling and requires thermal treatment at 700–1000 °C for a fairly long duration of 4–6 h (Bandara et al. 2016). Studies report that while neodymium extraction can be as high as 100%, dysprosium extraction could be in the range of 60–94% depending on the extraction technique, temperature, and time (Bandara et al. 2016; Rasheed et al. 2021; Tunsu 2018).

The material extraction intensities assumed in this study are based on the previous studies (Bandara et al. 2016), which indicate a 100% recovery yield with the least impurity for Nd and Pr, and about 94% extraction yield for Dy considering the extraction time limit of 48 h. Based on these assumptions, there is a difference between the quantum of the magnetic minerals used in initial manufacturing and that recovered post-recycling. Table 3 summarizes the quantities of the magnet-based RE metals (Nd, Pr, Dy) used initially for the manufacture of RAC compressors. Table 4 exhibits the quantities of the individual REEs recovered from scrap permanent magnets, via the two predominant recycling methods considered.

Table 3 Quantity of permanent magnet and rare earth materials used during the initial production of inverter RACs

Year of production	Share of inverter RAC	Inverter RAC units (in millions)	Magnet and rare earth used during production (tons)			
			Magnet	Nd	Dy	Pr
2015	10%	0.25	28.72	8.33	1.44	0.29
2016	13%	0.44	50.36	14.61	2.52	0.50
2017	18%	0.70	80.72	23.41	4.04	0.81
2018	24%	1.12	129.03	37.42	6.45	1.29
2019	33%	1.72	197.86	57.38	9.89	1.98
2020	44%	1.98	228.15	66.16	11.41	2.28

Table 4 Quantity of permanent magnet and rare earth materials recovered at the end of recycling process (100% collection and recycling of end-of-life RACs)

	Magnet and rare earth recovery via physical separation (tons)				Magnet and rare earth recovery via leaching method (tons)			
Year of end of life	Magnet	Nd	Dy	Pr	Magnet	Nd	Dy	Pr
2025	27.28	7.91	1.28	0.27	23.55	6.83	1.11	0.24
2026	47.85	13.88	2.25	4.78	41.30	11.98	1.94	0.41
2027	76.69	22.24	3.60	7.67	66.19	19.20	3.11	0.66
2028	122.58	35.55	5.76	12.26	105.80	30.68	4.97	1.06
2029	187.97	54.51	8.83	18.80	162.24	47.05	7.63	1.62
2030	216.74	62.86	10.19	21.67	187.08	54.25	8.79	1.87

Table 5 Different e-waste collection scenarios considered for sensitivity analysis

Scenario	Description
Business-as-usual (BAU) scenario	A 5% collection rate for the end-of-life RACs for the period up to 2030
Moderate collection rate (MCR) scenario	An improved e-waste collection scenario where the collection rate steadily increases at a CAGR of 1.43% to reach 30% by the end of 2030 from the current level of 5%
Accelerated formalized recycling (AFR) scenario	An aggressively enhanced e-waste collection scenario where the collection rate increases at a steep CAGR of 1.7% to reach 70% by the end of 2030 from the current level of 5%

Apart from the technical challenges associated with the recycling and recovery of crucial RE minerals, the lack of regulatory directives regarding the collection and commercial-scale recycling of electronic wastes is another hindering factor affecting the alternate source of critical materials. The total magnetic material recovery amounting to 680 tons and 586 tons via physical separation and leaching methods, respectively, assuming 100% formal e-waste collection and recycling between the years 2025 and 2030 will be quite difficult given the current state of affairs. Nevertheless, there had been amendments in India's E-Waste Management Rules in 2018 (PIB 2018), with a detailed mention of collection, storage, transportation, dismantling, recycling, and disposal of e-waste, as well as extended producer responsibility (EPR) (MoEFCC 2018), and a new draft has been drawn in 2022 aimed at improving the e-waste recycling in the country (PIB 2022). It is anticipated that mandatory provisions related to extended producer responsibilities (EPR) along with consumer awareness toward sustainability will improve the condition of formal recycling in the coming years.

In view of the recent regulatory changes, three different scenarios have been considered to reflect different possible levels of RAC collection and recycling. These three scenarios are summarized in Table 5, and they provide different trajectories for the EoL RAC collection and subsequent recycling for possible recovery of critical materials. This would help in gaining a more informed perspective on the scope of supply risk management through recycling and circular economy. Figures 8, 9, and 10 depict the magnet recovery for the three assumed different trajectories of RAC collection defined in Table 5.

Analysis of the three different e-waste collection rate scenarios highlights that if the e-waste collection and recycling situations do not change in the near future, then the *Business-as-usual* scenario will be able to cater to only 0.65% of the projected annual demand for NdFeB magnets in 2030. In the case of *Moderate collection rate (MCR)* scenario, upon improving the collection rate to 30% by 2030, the recycling route could help in meeting about 3% of the projected annual demand for magnets in 2030. In the third scenario of *Accelerated formalized recycling (AFR)* (70% e-waste recycling by 2030), recycling potential shows that the suggested alternate supply would be able to support about 9% of the projected annual demand for magnets in 2030. Tables 6, 7, and 8 show the recovered quantities of individual REEs from the

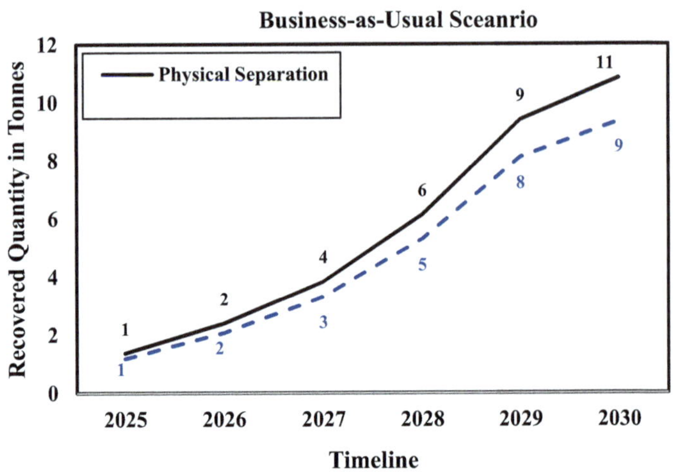

Fig. 8 Magnet recovery in business-as-usual (BAU) scenario via the two predominant recycling methods considered

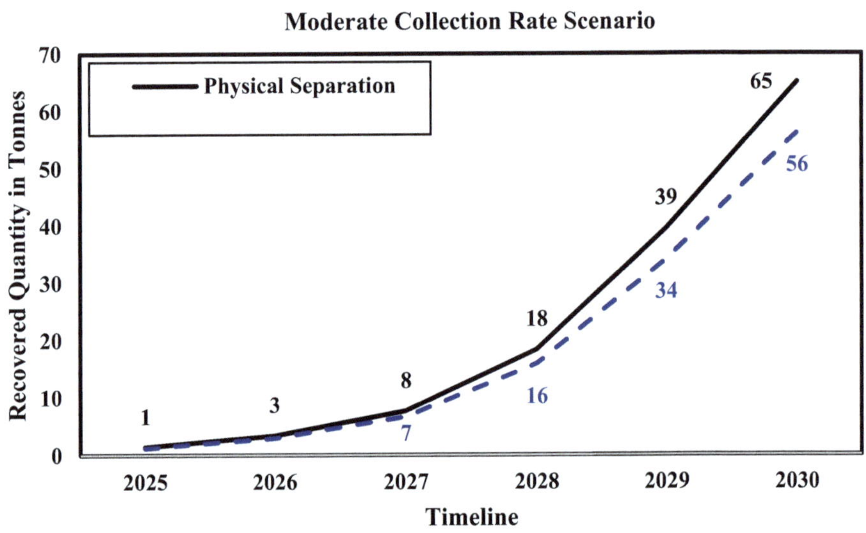

Fig. 9 Magnet recovery in moderate collection rate (MCR) scenario via the two predominant recycling methods considered

scrap magnets for the three different collection scenarios as well as two different recycling methodologies considered.

Figure 11 presents the summary of the conceptual template used in the study in the form of a schematic flow diagram (the dashed line indicates the scope of the present study). The different scenarios were chosen for assessing the possible variabilities that might be expected in the formal recovery and recycling of NdFeB

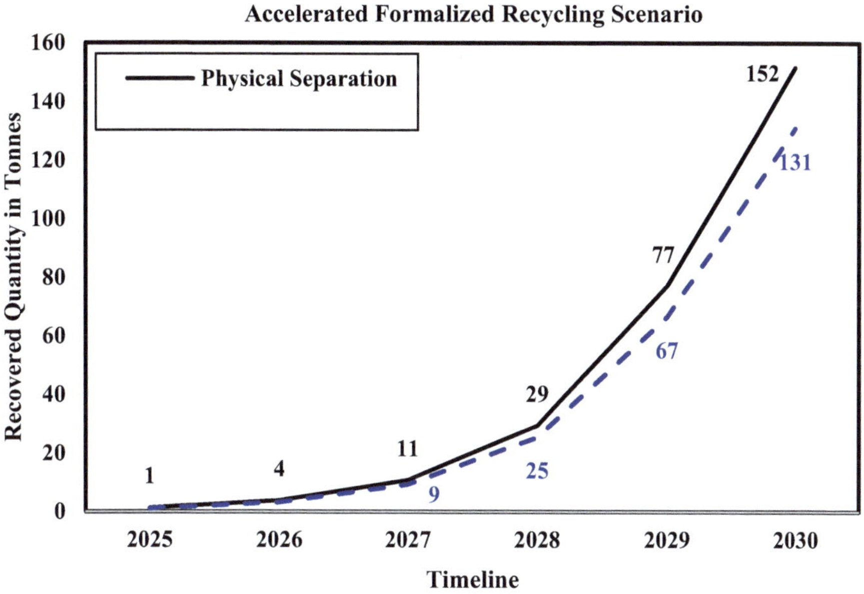

Fig. 10 Magnet recovery in accelerated formalized recycling (AFR) scenario via the two predominant recycling methods considered

Table 6 Quantity of permanent magnet and rare earth materials recovered in business-as-usual (BAU) scenario at the end of recycling process

Year of end of life	E-waste collection rate (in %) for recycling	No. of recycled inverter RACs (millions)	Magnet and rare earth recovery via physical separation (tons)				Magnet and rare earth recovery via leaching method (tons)			
			Magnet	Nd	Dy	Pr	Magnet	Nd	Dy	Pr
2025	5	0.01	1.36	0.40	0.06	0.014	1.18	0.34	0.06	0.01
2026	5	0.02	2.39	0.69	0.11	0.024	2.06	0.60	0.10	0.02
2027	5	0.04	3.83	1.11	0.18	0.038	3.31	0.96	0.16	0.03
2028	5	0.06	6.13	1.78	0.29	0.061	5.29	1.53	0.25	0.05
2029	5	0.09	9.40	2.73	0.44	0.094	8.11	2.35	0.38	0.08
2030	5	0.10	10.84	3.14	0.51	0.108	9.35	2.71	0.44	0.09

permanent magnets and associated rare earth elements from the RAC e-waste. An estimate of possible variability is important for planning purposes since they provide a baseline regarding the investment and infrastructure requirements toward building an alternate critical energy material supply chain.

In response to the limited availability and foreseeable scarcity of critical energy materials, capability building in terms of magnet and motor manufacturing and REE recycling is essential. The efficiency in material reuse would add substantially to the "greenness" of the envisaged transition to a cleaner and more energy-efficient future.

Table 7 Quantity of permanent magnet and rare earth materials recovered in moderate collection rate (MCR) scenario at the end of the recycling process

Year of end of life	E-waste collection rate (in %) for recycling	No. of recycled inverter RACs (millions)	Magnet and rare earth recovery via physical separation (tons)				Magnet and rare earth recovery via leaching method (tons)			
			Magnet	Nd	Dy	Pr	Magnet	Nd	Dy	Pr
2025	5	0.01	1.36	0.40	0.06	0.01	1.18	0.34	0.06	0.01
2026	7	0.03	3.35	0.97	0.16	0.03	2.89	0.84	0.14	0.03
2027	10	0.07	7.67	2.22	0.36	0.08	6.62	1.92	0.31	0.07
2028	15	0.17	18.39	5.33	0.86	0.18	15.87	4.60	0.75	0.16
2029	21	0.36	39.47	11.45	1.86	0.39	34.07	9.88	1.60	0.34
2030	30	0.60	65.02	18.86	3.06	0.65	56.13	16.28	2.64	0.56

Table 8 Quantity of permanent magnet and rare earth materials recovered in accelerated formalized recycling (AFR) scenario at the end of recycling process

Year of end of life	E-waste collection rate (in %) for recycling	No. of recycled inverter RACs (millions)	Magnet and rare earth recovery via physical separation (tons)				Magnet and rare earth recovery via leaching method (tons)			
			Magnet	Nd	Dy	Pr	Magnet	Nd	Dy	Pr
2025	5	0.01	1.36	0.40	0.06	0.01	1.18	0.34	0.06	0.01
2026	8	0.04	3.83	1.11	0.18	0.04	3.30	0.96	0.16	0.03
2027	14	0.10	10.74	3.11	0.50	0.11	9.27	2.69	0.44	0.09
2028	24	0.27	29.42	8.53	1.38	0.29	25.39	7.36	1.19	0.25
2029	41	0.71	77.07	22.35	3.62	0.77	66.52	19.29	3.13	0.67
2030	70	1.39	151.72	44.00	7.13	1.52	130.96	37.98	6.16	1.31

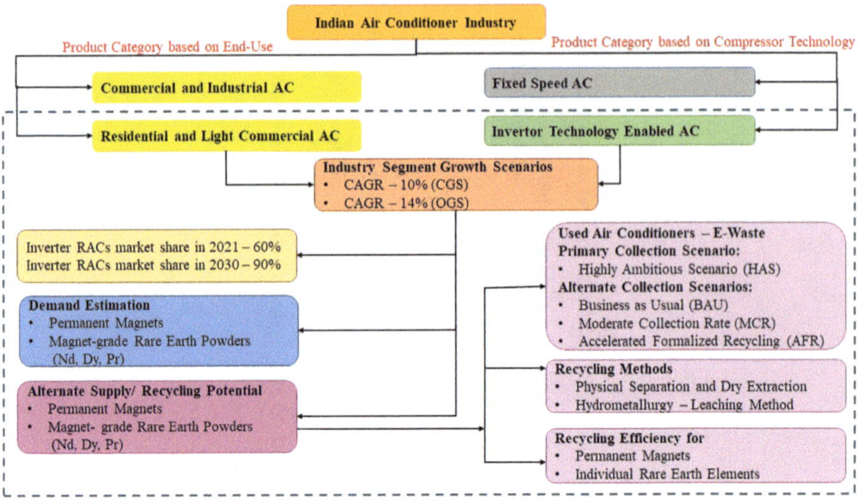

Fig. 11 Conceptual template for evaluating the potential for e-waste recycling for partial fulfillment of the demand for NdFeB magnet and magnet-based REEs in RAC market

By adopting a business-oriented approach toward building the industry ecosystem, India can enhance the potential for material recovery and recycling. However, for the viability of the product recycling route as an alternate source for meeting the critical material demand in the high-demand commercial durables (e.g., RACs), the collection rate of end-of-life products should improve substantially from the current level. Such innovative interventions would prove to be crucial in the coming decades when there will be a higher demand for residential air conditioners, and critical energy materials (e.g., Nd, Pr, Dy) will be short in supply. Therefore, it is essential to develop suitable strategies at the national level for attaining economies of scale in recycling of scrap magnets to make the recycling process economically viable. Further, the reprocessing facilities should be integrated with the e-waste collection and component segregation line to ensure seamless inflow of the scrap magnets from the EoL products.

Recently, Indian government has issued guidelines, promoting voluntary extended producer responsibility (EPR) to improve e-waste collection and recycling, accompanied by targets for minimum reuse of recovered minerals for the different subcomponent categories (PIB 2022). It is expected that in the near future, similar guidelines and policy interventions will be formulated for the effective collection of used electric motors and the recycling of critical components such as permanent magnets. This will further assist in building a robust indigenous manufacturing ecosystem, with enhanced production lines for highly energy-efficient products (as per the mandate of the Bureau of Energy Efficiency (BEE)), supported by the adoption of a sustainable 3R (recover, recycle, and reuse) model for material efficiency and also suitable financial arrangements.

5.1 *Reduction in Rare Earth Mining: Mitigation of Environmental Burden*

The principal sources of rare earth elements are *bastnaesite* (a fluorocarbonate commonly found in carbonatites and related igneous rocks), *xenotime* comprising yttrium phosphate (which predominantly occurs in mineral sand deposits), *loparite* (observed in alkaline igneous rocks), and *monazite* (a phosphate). In India, monazite is the principal source of REEs and thorium. The beach sand minerals (BSM) from placer deposits in India are a lean source of REEs (grade 0.06%). Therefore, a large volume of beach sand ore surface mining (~in excess of 50 lakh tons per annum) would be required to achieve a reasonable quantity of Nd-Pr compounds (~400–500 tons per annum). Large-scale beach sand mining may also attract illegal sand mining for construction activities. Sand mining results in reduced sediment availability, eventually leading to coastal erosion. Coastal erosion would cause loss of land, destruction of infrastructure, and displacement of coastal communities (Mohanty et al. 2023). Keeping this in mind, exploration for REEs is being carried out by the Geological Survey of India (GSI) and the State Directorates of Geology & Mining for alternate sources. The reconnaissance stage investigation (G-4) has identified a few potential inland areas for REEs and other minerals. However, a detailed assessment of the grade of the ore in a case-specific manner would be crucial, since lean ore would lead to large-volume mining, leading to a higher environmental burden. Many identified locations in Northeast India also have biological hotspots in the vicinity. The sensitive and fragile ecosystems can potentially face dire consequences in case imprudent and unabated mining activities take place in those areas.

There are two primary methods for REE mining. The first method involves the removal of topsoil and the creation of a leaching pond where chemicals are added to the extracted ore to separate metals. However, toxic chemicals from such leaching ponds may leak into groundwater aquifers leading to devastating effects on the ecology and the environment. The second method involves the drilling of holes into the ground and the use of polyvinyl chloride (PVC) pipes and rubber hoses to pump chemicals into the earth for the extraction of desired metals. This method also creates a leaching pond with similar problems. Sometimes, the PVC pipes are not removed, leading to the natural aging of plastic material and the incursion of microplastics into the groundwater. Both methods produce a substantial amount of toxic waste, with a high risk of environmental and health hazards. For every ton of RE metal produced from the hard igneous rock deposits, the mining process yields roughly 13 kg of toxic dust, 9600–12,000 cubic meters of waste gas, 75 cubic meters of wastewater, and 1 ton of radioactive residue. This may however vary from one place to another based on the geological attributes and the type of deposit being exploited. This indicates that mining activities have the potential to contaminate air, water, and soil simultaneously. Looking at these concerns, the recovery and recycling of REEs provide some cushion for developmental activities when observed from the environmental impact point of view.

6 Conclusions

This study presents two different growth scenarios for the energy-efficient residential air conditioners (conservative and optimistic growth scenarios, respectively) to reflect the possible future demand for the NdFeB magnets and the critical REEs (Nd, Pr, Dy). From a thorough scrutiny of available literature, it is identified that the heavy consumer durable market in India (including RACs) could face the supply risk of NdFeB permanent magnets used in energy-efficient end products. This is primarily due to the limited availability of critical magnet-based rare earth materials—neodymium, praseodymium, and dysprosium. A price shock may also deter the growth of the consumer durable industry. Considering India's current status in the manufacturing of air conditioner compressors and the untapped market opportunities related to end-of-life electronic products, a conceptual template comprising recycling and remanufacturing has been suggested. Such an approach can reduce the country's import dependency for the critical materials and components used in the subsystems and systems of the commercially important product.

Currently, recovery and recycling are driven primarily by environmental needs, focused on reducing the toxicity attributed to e-waste disposal. A small fraction of the total e-waste (i.e., a few specific components that generate value for the dealers) reaches the formalized recycling facilities. Therefore, there is ample opportunity to tap the recycling potential provided suitable financial arrangements and formal ties can be worked out among the various parts of the waste collection and waste recycling value chains, leading to an integrated supply chain. Further, global advancements in the technical aspects of recycling should be closely followed, and necessary technological improvements should be imbibed and indigenized to enhance sustainable manufacturing ecosystem. By these means, the pressure on the extraction of virgin raw material through energy- and emission-intensive mining processes can be reduced by a substantial amount in the coming decades. Moreover, recent research related to electric motor design and specification supports the application of smaller but multiple rare earth-based permanent magnets for improved magnetic and coercivity properties, rather than big, centralized ones. In view of these developments, recycled magnets of smaller size would prove to be very useful for the end-product manufacturers as well as the makers of the magnets and motors. Apart from the initial support from the government for the creation of facilities and capacities, these trends would likely attract a lot of interest from business enterprises and original equipment manufacturers to invest and establish modern recycling facilities which could benefit the economy in the form of alternate supply chain and reduced burden on the environment.

Acknowledgment The corresponding author (Rudrodip Majumdar) is highly grateful to Dr. Shailesh Nayak (Director, NIAS) for his invaluable insights and comments, which helped in enhancing the quality of the chapter.

References

AIEEE (2021) Cooling is the new hot: transitioning India to super energy-efficient room air conditioners. Alliance Energy Efficient Econ. Retrieved from https://aeee.in/cooling-is-the-new-hot/. Last Accessed 5 Feb 2023

Alonso E, Sherman AM, Wallington TJ, Everson MP, Field FR, Roth R, Kirchain RE (2020) Evaluating rare earth element availability: a case with revolutionary demand from clean technologies. Environ Sci Technol 46(6):3406–3414. https://doi.org/10.1021/es203518d. Erratum in: Environ Sci Technol 2012 Apr 17;46(8):4684

Alves Dias P, Bobba S, Carrara S, Plazzotta B (2020) The role of rare earth elements in wind energy and electric mobility. Publication Office of the European Union, Luxembourg., ISBN 978-92-79-27016-4. https://doi.org/10.2760/303258

Baba K, Hiroshige Y, Nemoto T (2013) Rare earth magnet recycling. Hitachi Rev 62(8) Retrieved from https://www.hitachi.com/rev/pdf/2013/r2013_08_105.pdf. Last Accessed 5 Feb 2023

Bandara HMD, Fielda KD, Emmert MH (2016) Rare earth recovery from end-of-life motors employing green chemistry design principles. Green Chem 18:753–759. Royal Society of Chemistry. https://doi.org/10.1039/C5GC01255D

Barteková E (2016) De Lima IB, Filho WL (eds) The role of rare earth supply risk in low-carbon technology innovation, chapter 10 in rare earths industry—technological, economic, and environmental implications. Elsevier, pp 153–169. ISBN 9780128023280. https://doi.org/10.1016/B978-0-12-802328-0.00010-3

BEE (2006) Standards and labelling program. Bureau Energy Efficiency. Retrieved from http://beeindia.gov.in/en. Last Accessed 5 Feb 2023

BEE (2020a) Light commercial air conditioners (02 March 2020). Schedule 24. Bureau of Energy Efficiency. Retrieved from Schedule_LCAC.pdf (beeindia.gov.in). Last Accessed 5 Feb 2023

BEE (2020b) Room air conditioners. schedule—3(A). Bureau Energy Efficiency. Retrieved from Microsoft Word—Schedule Annexure-3A_Rev V 2_.docx (beeindia.gov.in). Last Accessed 5 Feb 2023

M. Brain, C.W. Bryant and S. Elliott (2011). How air conditioners work. Retrieved from https://home.howstuffworks.com/ac2.htm. Last Accessed 5 Feb 2023

Bunting E-Magnets (2023) NdFeB magnet grades. Retrieved from https://tinyurl.com/mry2876s. Last Accessed 5 Feb 2023

Campbell P (2008) System cost analysis for an interior permanent magnet motor. AMES Laboratory (Creating Materials and Energy Solutions), US Department of Energy. Retrieved from https://www.osti.gov/servlets/purl/940187. Last Accessed 5 Feb 2023

Chen IY, Chen YM, Chang YJ, Wei CS, Wang CC (2009) A comparative study between a constant-speed air-conditioner and a variable-speed air-conditioner. ASHRAE Trans 115(1):326–332. Retrieved from https://tinyurl.com/yc5uaz2w. Last Accessed 5 Feb 2023

Collocott SJ, Dunlop JB, Gwan PB, Kalan BA, Lovatt HC, Wu W (2004) Applications of rare-earth permanent magnets in electrical machines: from motors for niche applications to hybrid electric vehicles. Conference: 2004 China Magnetic Materials and Devices Assoc.

Coney MW, Stephenson P, Malmgren A, Linnemann C, Morgan RE (2002) Development of a reciprocating compressor using water injection to achieve quasi-isothermal compression. Int Comp Eng Conf. Retrieved from https://docs.lib.purdue.edu/icec/1508/

Constantinides S (2012) The demand for rare earth materials in permanent magnets. 51st annual conference of metallurgists. Retrieved from https://tinyurl.com/yc7e5tnv. Last Accessed 5 Feb 2023

Constantinides S (2021) Industrial considerations for the recycling of RE magnets. Circularizing rare earth elements in magnet applications. Critical Minerals Institute. Retrieved from https://tinyurl.com/yfwttb56. Last Accessed 5 Feb 2023

Daikin (2023) Inverter for energy savings. Retrieved from https://www.daikin.com/corporate/why_daikin/benefits/inverter. Last Accessed 5 Feb 2023

Davis LW, Gertler PJ (2015) Contribution of air conditioning adoption to future energy use under global warming. Econ Sci 112(19):5962–5967. https://doi.org/10.1073/pnas.1423558112
De Campos MF, De Castro JA (2020) Current trends in recycling, usage and market of rare-earths. Conference paper: sustainable industrial processing summit and exhibition
De Cian E, Pavanello F, Randazzo T, Mistry MN, Davide M (2019) Households' adaptation in a warming climate- air conditioning and thermal insulation choices. Environ Sci Policy 100:136–157. ISSN 1462-9011. https://doi.org/10.1016/j.envsci.2019.06.015
Dent PC (2012) Rare earth elements and permanent magnets. J Appl Phys 111:07A721. https://doi.org/10.1063/1.3676616
Dong Y, Coleman M, Miller SA (2021) Greenhouse gas emissions from air conditioning and refrigeration service expansion in developing countries. Annu Rev Env Resour 46(1):59–83. https://doi.org/10.1146/annurev-environ-012220-034103
Dubas F, Espanet C, Miraoui A (2005) Design of a high-speed permanent magnet motor for the drive of a fuel cell air-compressor. IEEE Vehicle Power Propul Conf. https://doi.org/10.1109/VPPC.2005.1554621
Dutt A, Tyagi A (2022) Building resilient mineral supply chains for energy security. Council on Energy, Environment and Water, New Delhi. Retrieved from https://tinyurl.com/2xvvxc52. Last Accessed 5 Feb 2023
Eggert R, Wadia C, Anderson C, Bauer D, Fields F, Meinert L, Taylor P (2016) Rare earths: market disruption, innovation, and global supply chains. Annu Rev Env Resour 41(1):199–222. https://doi.org/10.1146/annurev-environ-110615-085700
Energy Conservation Act, 2001. Government of India. Retrieved from https://legislative.gov.in/sites/default/files/A2001-52.pdf. Last Accessed 5 Feb 2023
European Commission (2017) REE4EU: integrated high temperature electrolysis (HTE) and ion liquid extraction (ILE) for a strong and independent European Rare Earth Elements Supply Chain. Retrieved from https://cordis.europa.eu/project/id/680507. Last Accessed 5 Feb 2023
Fastenau RHJ, van Loenen EJ (1996) Applications of rare earth permanent magnets. J Magn Magn Mater 157–158:1–6. ISSN 0304-8853. https://doi.org/10.1016/0304-8853(95)01279-6
Ferreira G, Critelli J (2022) China's global monopoly on rare-earth elements. Parameters 52(1):57–72. https://doi.org/10.55540/0031-1723.3129
Fountain H, Rojanasakul M (2023) The last 8 years were the hottest on record. The New York Times. Retrieved from https://tinyurl.com/3snm93ua. Last Accessed 5 Feb 2023
Fujita Y, McCall SK, Ginosar D (2022) Recycling rare earths: perspectives and recent advances. MRS Bull 47:283–288. https://doi.org/10.1557/s43577-022-00301-w
Gieras JF (2010) Permanent magnet motor technology—design and applications. Taylor and Francis Group
Grand View Research (2022) Global HVAC systems market size reports 2022–2030. Retrieved from https://tinyurl.com/yesk7rbs. Last Accessed 5 Feb 2023
Gschneidner Jr KA (2015) The rare earth crisis—the supply/demand situation for 2010–2015. Mater Matters 6(2):32–37. Retrieved from https://tinyurl.com/36fxd8jw
Gupta V, Ganesan K (2014) India's critical mineral resources: a trade and economic analysis. Council on Energy, Environment and Water, New Delhi. Retrieved from https://tinyurl.com/mrymj5by. Last Accessed 5 Feb 2023
Habib K, Wenzel H (2014) Exploring rare earths supply constraints for the emerging clean energy technologies and the role of recycling. J Clean Prod 84:348–359. ISSN 0959-6526. https://doi.org/10.1016/j.jclepro.2014.04.035
IEA (2018) The future of cooling. International Energy Agency (IEA). Retrieved from The Future of Cooling—Analysis—IEA. Last Accessed 5 Feb 2023
IEA (2021) The role of critical minerals in clean energy transitions. International Energy Agency (IEA). Retrieved from https://www.iea.org/reports/the-role-of-critical-minerals-in-clean-energy-transitions. Last Accessed 5 Feb 2023
IEA (2022a) Cooling—fuels and technologies. International Energy Agency (IEA) Retrieved from Cooling—Fuels & Technologies—IEA. Last accessed 5 Feb 2023

IEA (2022b) Space cooling. International Energy Agency (IEA). Retrieved from https://www.iea.org/reports/space-cooling. Last Accessed 5 Feb 2023

Im SY, Park SH, Kim JH, Chin JW, Cha KS, Lim MS (2021) Temperature prediction of ultra-high-speed SPMSM for FCEV air compressor considering PWM current harmonics. IEEE vehicle power and propulsion conference (VPPC)-2021, Gijon, Spain, pp 1–6, https://doi.org/10.1109/VPPC53923.2021.9699318

IMA (2018) Applications of neodymium magnets in electric motors. IMA-magnet factory and magnetic applications. Retrieved from https://tinyurl.com/4wtmjrpk. Last Accessed 5 Feb 2023

Indian Bureau of Mines (2016) Indian Minerals Yearbook 2014 (Part-III: Mineral Reviews), 53rd edn., Rare Earths (Advance Release). Retrieved from https://tinyurl.com/57jzf6yr. Last Accessed 5 Feb 2023

IPCC (2021) (Working Group-I Report—Climate Change 2021: the physical science basis. Intergovernmental Panel on Climate Change (IPCC) Retrieved from https://tinyurl.com/mtty54s6. Last Accessed 5 Feb 2023

IPCC (2022) Working group-II report. Climate change 2022: impacts, adaptation and vulnerability. Intergovernmental Panel on Climate Change (IPCC). Retrieved from https://www.ipcc.ch/report/ar6/wg2/. Last Accessed 5 Feb 2023

Jain A (2022) What is inverter AC and how it is different from non-inverter AC? Bijli Bachao. October 2022. Retrieved from https://tinyurl.com/s5h4xh7e. Last Accessed 5 Feb 2023

JRAIA (2019) World air conditioner demand by region, June 2019. Japan Refrigeration and Air Conditioning Industry Association (JRAIA). Retrieved from https://www.jraia.or.jp/english/World_AC_Demand.pdf. Last Accessed 5 Feb 2023

Kanchwala H (2022) Whirlpool AC in India—review 2022 (Bijli Bachao). Retrieved from https://tinyurl.com/4mj7na4r. Last Accessed 5 Feb 2023

Karali N, Shah N, Park WY, Khanna N, Ding C, Lin J, Zhou N (2020) Improving the energy efficiency of room air conditioners in China: costs and benefits. Appl Energy 258:114023. ISSN 0306-2619. https://doi.org/10.1016/j.apenergy.2019.114023

Karkour S, Ihara T, Kuwayama T, Yamaguchi K, Itsubo N (2021) Life cycle assessment of residential air conditioners considering the benefits of their use: a case study in Indonesia. Energies 14(2):447. https://doi.org/10.3390/en14020447

Khosla R, Agrawal A, Sircar N, Chatterjee D (2021) The what, why, and how of changing cooling energy consumption in India's urban households. Environ Res Lett 16(4):044035. https://doi.org/10.1088/1748-9326/abecbc

Kim YK, Lee J (2005) A comparative study of high-speed permanent magnet synchronous motor for air compressor. IEEE international magnetics conference (INTERMAG), Nagoya, pp 663–664, https://doi.org/10.1109/INTMAG.2005.1463760

Lu D (2019) Top countries that manufacture HVACR components. GSI. Retrieved from https://tinyurl.com/y4deahmz. Last Accessed 5 Feb 2023

Ma D, Henderson J (2021) The impermanence of permanent magnets: a case study on industry, Chinese production, and supply constraints. Retrieved from https://tinyurl.com/bdxf7uu6. Last Accessed 5 Feb 2023

Mishra A, Agrawal P, Srivastava SP (2014) A comprehensive analysis and implementation of vector control of permanent magnet synchronous motor. Int J Power Energy Convers 5:1–23. https://doi.org/10.1504/IJPEC.2014.059982

MoEFCC (2018) Notification dt. 22 March 2018, Ministry of Environment, Forest and Climate Change (MOEF&CC), GOI. Retrieved from https://tinyurl.com/2p8yyd5n. Last Accessed 5 Feb 2023

MoEFCC (2019) India cooling action plan. Ministry of Environment, Forest, and Climate Change (MOEF&CC), Government of India. Retrieved from https://tinyurl.com/26ddkad2. Last Accessed 5 Feb 2023

Mohanty A, Rajani MB, Majumdar R, Nayak S (2023) Improved geospatial analysis of shoreline modification using a weighted-average-based novel formulation. Earth Surf Process Landf 48(5):863–886. https://doi.org/10.1002/esp.5522

MoHUA (2016). Building Bye Laws, 2016. Ministry of Urban and Housing Affairs Development, Government of India. Retrieved from https://tinyurl.com/4c22bxpr. Last Accessed 5 Feb 2023

Mordor Intelligence (2022) Air conditioner market—growth, trends, covid-19 impact, and forecasts (2023–2028). Retrieved from https://tinyurl.com/3cbayyp5. Last Accessed 5 Feb 2023

Mukherjee W (2022) LG electronics becomes the leader in dual inverter AC market. Econ Times Bureau. Retrieved from https://tinyurl.com/2ths4huk. Last Accessed 5 Feb 2023

Ohyama K, Kondo T (2008) Energy-saving technologies for inverter air conditioners. IEEJ Trans Electr Electron Eng. https://doi.org/10.1002/tee.20254

Ormerod J (2022) Croat J, Ormerod J (eds) Permanent magnet markets and applications. Chapter 12 in modern permanent magnets. Woodhead publishing series in electronic and optical materials, pp 403–434. ISBN 9780323886581. https://doi.org/10.1016/B978-0-323-88658-1.00012-1

Motilal Oswal (2017a) Report on room air conditioners—focus shift to inverters (2017). Retrieved from https://tinyurl.com/yp8w45p9. Last Accessed 5 Feb 2023

Motilal Oswal (2017b) Report on room air conditioners. Transition to inverters gathers pace. Retrieved from https://tinyurl.com/4b6acymk. Last Accessed 5 Feb 2023

Motilal Oswal (2017c) Report on capital goods—focus on upcoming summer season; inverter models gather pace. Retrieved from https://tinyurl.com/563kk4bv. Last Accessed 5 Feb 2023

Motilal Oswal (2018a) Report on room air conditioners—update on consumer durable sector. Retrieved from https://tinyurl.com/6dssbw4e. Last Accessed 5 Feb 2023

Motilal Oswal (2018b) Report on room air conditioners. Inverters the key focus category in room air conditioners. Retrieved from https://tinyurl.com/bdstr3km. Last Accessed 5 Feb 2023

Motilal Oswal (2018c) Report on room air conditioners. Inventory in channel high post weak 1QFY19 sales. Retrieved from https://tinyurl.com/uj43sy2a. Last Accessed 5 Feb 2023

Palmer C (2022) The drive for electric motor innovation. Engineering 8:9–11. https://doi.org/10.1016/j.eng.2021.11.007

Pandita S, Kumar PVNK, Walia A, Ashwin TP (2020) Policy measures and impact on the market for the room air conditioners in India. CLASP India Bureau Energy Efficiency. Retrieved from https://tinyurl.com/2uyc5m8h. Last Accessed 5 Feb 2023

Parkin B (2021) Indian demand for air-conditioning heats up climate fears. Fin Times. Retrieved from https://tinyurl.com/3xt2scx5. Last Accessed 5 Feb 2023

Pavanello F, De Cian E, Davide M, Mistry M, Cruz T, Bezerra P et al (2021) Air-conditioning and the adaptation cooling deficit in emerging economies. Nat Commun 12(1):6460. https://doi.org/10.1038/s41467-021-26592-2

PIB (2018) E-Waste Management Rules amended for effective management of E-Waste in the country. Press Information Bureau (PIB), Ministry of Environment, Forest and Climate Change (MoEF&CC), GOI. Retrieved from https://pib.gov.in/newsite/PrintRelease.aspx?relid=177949. Last Accessed 5 Feb 2023

PIB (2020) BEE notifies new energy performance standards for air conditioners (2020). Press India Bureau (PIB). Ministry of Power (MoP), Govt' of India. Retrieved from https://pib.gov.in/PressReleasePage.aspx?PRID=1598508. Last Accessed 5 Feb 2023

PIB (2022) Re-cycling of E-waste (December 2022). Press Information Bureau (PIB), Ministry of Environment, Forest and Climate Change (MOEF&CC), GOI. Retrieved from https://pib.gov.in/PressReleasePage.aspx?PRID=1881761. Last Accessed 5 Feb 2023

Rasheed MZ, Song M, Park S-m S, Nam S, Hussain J, Kim TS (2021) Rare earth magnet recycling and materialization for a circular economy—a Korean perspective. Appl Sci 11(15) Article 6739. https://doi.org/10.3390/app11156739

Semones B (1985) Volumetric improvements in high energy magnet motors. IEEE Trans Magn 21(5):1948–1951. https://doi.org/10.1109/TMAG.1985.1064063

Shi K, Chen Y, He Z, Wang J, Li Y (2019) Design of permanent magnet synchronous motor control system for electric vehicle air conditioning compressor based on vector control. Open Access Lib J 6:1–9. https://doi.org/10.4236/oalib.1105157

Shinetsu Rare Earth Magnets (2007). Retrieved from https://www.shinetsu-rare-earth-magnet.jp/e/rare/. Last Accessed 5 Feb 2023

Siriwardhana M, Namal DDA (2017) Comparison of energy consumption between a standard air conditioner and an inverter-type air conditioner operating in an office building. SLEMA J 20(1–2):1–6. https://doi.org/10.4038/slemaj.v20i1-2.5

Statista (2016a) Demand for rare earths in consumer electronic permanent magnets worldwide from 2010 to 2025. Retrieved from https://tinyurl.com/4fafv4c9. Last Accessed 5 Feb 2023

Statista (2016b) Demand of rare earths in industrial permanent magnets worldwide from 2010 to 2025 (in metric tons). Retrieved from https://tinyurl.com/3xskh395. Last Accessed 5 Feb 2023

Statista (2018). Share of households that have air-conditioning (AC) worldwide in 2016, by country. Retrieved from https://tinyurl.com/3v7bp9wk. Last Accessed 5 Feb 2023

Statista (2023a) Distribution of rare earth element consumption worldwide in 2021, by end-use. Retrieved from https://tinyurl.com/3d983w7b. Accessed 5 Feb 2023

Statista (2023b) Distribution of rare earths production worldwide as of 2022, by country. Retrieved from https://tinyurl.com/mw7v49p3. Last Accessed 5 Feb 2023

Texas Comptroller (2021) Rare earth elements supply chain. Retrieved from Rare Earth Elements Supply Chain (texas.gov). Last Accessed 5 Feb 2023

The Hindu Business Line (2018) LG India switches to inverter ACs. Retrieved from https://tinyurl.com/4f9sdtwr. Last Accessed 5 Feb 2023

The Print (2021) Rare earths, their strategic significance, China's monopoly & why it matters to the Quad Retrieved from https://tinyurl.com/4xjfk665. Last Accessed 5 Feb 2023

The Royal Society (2022) The basics of climate change. Retrieved from https://tinyurl.com/yub6jw97. Last Accessed 5 Feb 2023

Trench A, Sykes JP (2020) Rare earth permanent magnets and their place in the future economy. Engineering 6(2):115–118., ISSN 2095-8099. https://doi.org/10.1016/j.eng.2019.12.007

Tuholske C, Caylor K, Funk C, Evans T (2021) Global urban population exposure to extreme heat. Soc Sci 118(41). https://doi.org/10.1073/pnas.2024792118

Tunsu C (2018) Hydrometallurgy in the recycling of spent NdFeB permanent magnets, chapter 8 in waste electrical and electronic equipment recycling. Woodhead Publ Ser Electron Opt Mater:175–211. ISBN 9780081020579. https://doi.org/10.1016/B978-0-08-102057-9.00008-1

USGS (2020) Rare earth data sheet—minerals commodity summaries. United States Geological Survey (USGS) Retrieved from https://pubs.usgs.gov/periodicals/mcs2020/mcs2020-rare-earths.pdf. Last Accessed 5 Feb 2023

Uzhegov N, Smirnov A, Park CH, Ahn JH, Heikkinen J, Pyrhönen J (2017) Design aspects of high-speed electrical machines with active magnetic bearings for compressor applications. IEEE Trans Ind Electron 64(11):8427–8436. https://doi.org/10.1109/TIE.2017.2698408

World Meteorological Organization (WMO) (2022) Provisional state of the global climate. Retrieved from https://tinyurl.com/yhbyp9dk. Accessed 5 Feb 2023

Yang Y, Walton A, Sheridan R, Güth K, Gauß R, Gutfleisch O, Buchert M, Steenari BM, Gerven TV, Jones PT, Binnemans K (2017) REE recovery from end-of-life NdFeB permanent magnet scrap: a critical review. J Sustain Metall 3:122–149. https://doi.org/10.1007/s40831-016-0090-4

Yun GY, Steemers K (2011) Behavioural, physical and socio-economic factors in household cooling energy consumption. Appl Energy 88(6):2191–2200. ISSN 0306-2619. https://doi.org/10.1016/j.apenergy.2011.01.010

Environmental Impacts and Government Policies for Responsible Management of E-Waste

Nidhi Pandey and Pankaj Pathak

Abstract The exponential advancement of technology, along with our demanding consumption habits, has ended up in an alarming growth in electronic waste (e-waste) creation, posing a severe environmental risk. This chapter highlights the environmental implications of e-waste as well as government attempts to promote responsible waste management, sustainable consumption, and separation of raw materials from these wastes. Dangerous chemicals, including Hg, Pb, and Cd, along with brominated flame retardants (BFRs) are released into the atmosphere during illegal disposal or informal recycling of e-waste, eventually building up in our ecosystem and disrupting the normal ecological cycle. It results in heavy metal contamination in the soil and water, emissions in the air, as well as negative impacts on human lives such as reproductive abnormalities and respiratory disease. The government policy like extended producer responsibility (EPR) implementation frameworks emphasize the collection and recycling systems and promote the sustainable waste management practices. EPR requires producers to take responsibility for their goods complete life cycle, including proper disposal and recycling. Collaboration among governments, manufacturers, consumers, and recycling sectors is required to manage e-waste effectively. Thus, environmental risks can be reduced by adopting ecosystem-friendly practices in sustainable e-waste management and nurturing a circular economy.

Keywords E-waste · Hazards · Recycling · Circular economy · Sustainable resources

N. Pandey · P. Pathak (✉)
Resource Management Lab, Department of Environmental Science & Engineering, SRM University Andhra Pradesh, Guntur, India

P. Pathak et al. (eds.), *Anthropogenic Environmental Hazards*,
https://doi.org/10.1007/978-3-031-41013-0_4

1 Introduction

Incessant demands of electrical and electronics, including white goods, ICT equipment, and corporate and residential items, have made our life easy and comfortable. However, after reaching its end of life (EoL), it has been thrown as waste, which is called e-waste (Pathak et al. 2017). E-waste is described as a heterogeneous mixture of ferrous and nonferrous metals, including ceramics and plastic components, by the Association of Plastics Manufacturers in Europe (APME) (Srivastava and Pathak 2019). Commonly, household e-waste is disposed of along with municipal solid waste to the landfill or sold to vendors (authorized or unauthorized), and further, it reaches landfills (Ikhlayel 2018).

Globally, e-waste production reached 53.6 Mt in 2019, and only 17% of this was managed in a sustainable way that recover $9.4 billion in raw materials including iron, gold, copper, and other precious raw minerals (Forti et al. 2020). As a result, the fate of about 83% of the generated e-waste remains unknown or unaccounted for. With a possible loss of $47.6 billion in precious metals, garbage may be handled and repurposed informally or abandoned, burned, exchanged, or even recycled on occasion (Baldé et al. 2022). It is predicted that the generation of e-waste will increase by 74.7 Mt in 2030 and up to 110 Mt in 2050 (Tabelin et al. 2021). E-waste contains valuable metals and minerals such as copper, cobalt, gold, and platinum, as well as potential environmental concerns such as polychlorinated biphenyls (PCBs) and polybrominated diphenyl ethers (PBDEs) (Li and Achal 2020; Pathak and Pandey 2023).

Burning electronic debris can result in the release of dioxins, furans, polycyclic aromatic hydrocarbons (PAHs), polyhalogenated aromatic hydrocarbons (PHAHs), and hydrogen chloride (Das et al. 2021; Pathak et al. 2023). Further, informal recycling of e-waste that recover valuable metals but has a negative ecological impact is not a sustainable process. Moreover, developed nations export their unspecified amount of e-waste to developing nations that goes for informal recycling, which violates the Basel Agreement because recycling methods incorporate igniting and dissolving into deep acids without any effort taken to safeguard public's well-being or the surroundings (Kumar et al. 2017).

This type of recycling causes significant localized pollution at first, followed by the toxins spreading into receiving streams and food networks. Toxins are often introduced to the general public through a variety of avenues, including smoke, dust, drinking water, and food. However, because they handle electronic debris, e-waste employees are at a higher risk due to direct skin contact and inhalation. Furthermore, the influence of e-waste might extend beyond its immediate domain, as certain manufactured or agricultural items destined for export may contain contaminants linked with e-waste inadvertently. To determine the aforementioned global generation of e-trash, data on the number attributed to objects in use is required. Since these statistics are usually available in wealthy countries, Eq. 1 can be used to estimate how much e-waste is produced:

$$E = \frac{MN}{L} \tag{1}$$

The addition of a given product to the yearly creation of electronic trash, E (kg/year), is determined by its mass (M, kg) and the number of operational units (N) including its typical lifespan (L, years).

India holds the top ten position for the generation of e-waste in the world and is facing a severe problem amid the long-term disposal of such waste. As a result, the Indian government has been working for several years to provide an established foundation and judicial structure for our country's e-waste management. E-waste's massive volume and intricate composition, which contain lethal chemicals and toxic metals, endanger both the human health and the environment. E-waste contains thousands of distinct compounds, many of which are classified as hazardous or radioactive waste and vary depending on the products/devices generated. In mammals, the main physical repercussions include abnormalities of the brain, lungs, thyroid, and reproductive systems (Pathak et al. 2017). E-waste disposal releases dangerous heavy metals and poisonous gases which may eventually contaminate groundwater (Zhang et al. 2022). Groundwater contamination from metal deposits in the soil, including As, Cd, Cu, Pb, Se, and Zn, poses a persistent threat to the environment and human health. Lack of health and safety laws and inadequate recycling procedures, like landfill discarding, disassembling, improper cutting, ignition, and acid discharge, increase the threat to the people handling e-waste and the surrounding environment (Pathak et al. 2019; Pathak and Pandey 2023).

Further, heavy metal concentrations were found in elevated amounts in natural plant specimens (*Cynodon dactylon*) from e-waste reprocessing sectors. Studies on threat estimation based on thorough metal concentration and multivariate analysis have shown considerable discrepancies across sample locations and solid proof of contamination with the heavy metal as a result of nonformal e-waste reprocessing. Long-term nonformal reprocessing of e-waste could result in elevated quantities of dangerous heavy metals accumulating in the upper soil layer, vegetation, as well as groundwater, posing a high risk to both the environment and the personnel. This needs the immediate deployment of corrective procedures to lessen heavy metal pollutants in e-waste reprocessing operations.

2 The Global Framework of E-Waste Generation

Different continents in the world generated 53.6 Mt of e-waste in 2019 as shown in Fig. 1. Production of electronic waste worldwide has increased by 9.2 Mt since 2014, and by 2030, it is projected to reach 74.7 Mt, nearly doubling in just 16 years. The main drivers for the e-waste generation rate are higher usage rates of electrical and electronic equipment (EEEs), short service lives, and limited repair choices of e-wastes (Forti et al. 2020).

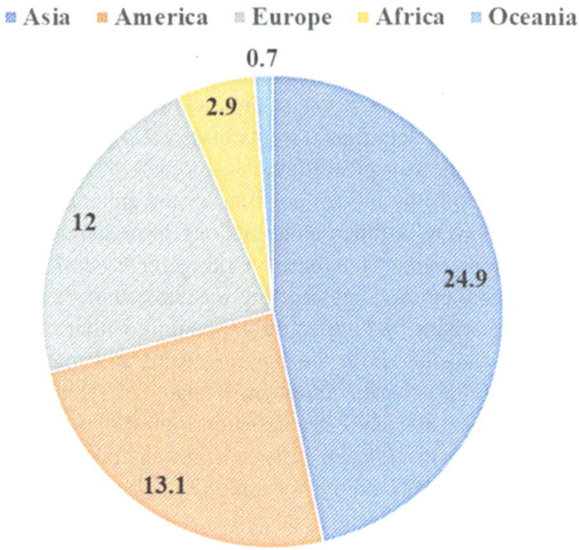

Fig. 1 E-waste generated by different continents in the year 2019 (in Mt)

The international approach helps to harmonize the measurement framework and indicators to create an integrated and comparable global measuring framework for e-waste. The framework quantifies and identifies the essential elements of a country's e-waste. The following indications can be generated using the framework:

1. The total amount of e-waste that has been commercialized (placed on the market (POM), kg/person). This illustrates how large the domestic e-goods market is and the overall e-waste produced (kg/person). This indicates the amount of electronic garbage generated on a national scale.
2. E-waste that has been formally collected (in kg/person). The quantity of electronic waste that is formally collected using the collection mechanism is as follows:

$$\text{E} - \text{waste collection center} = \frac{\text{Total e} - \text{waste recycled}}{\text{Total e} - \text{waste generated}} \times 100 \quad (2)$$

3. This metric indicates how successfully the government's collection procedures are functioning (Forti et al. 2020).

In the region's low- and middle-income economies, the possession rate of large equipment and temperature exchange equipment is frequently no more than two appliances per home. These large, bulky appliances have relatively high unit weights and long lifespans. In contrast, the unit weight of tiny equipment is often lower. Smaller equipment is usually abandoned because it is purchased more frequently and has a shorter lifespan.

2.1 Mass Generation of E-Waste by Different Regions of India

Countries like India are battling an exponentially growing amount of e-waste produced domestically or illicitly. The average annual growth rate of electronic waste is 3–5% worldwide (Arya and Kumar 2020). The top nine states of India contribute ~70% of the total e-waste as shown in Table 1. It has been seen that e-waste is expanding within developing nations due to the insufficient imposition of existent legislation. Trading, repairing, and extracting resources from discarded electronic equipment have developed a new economic sector. Despite providing inception of an income in favor of the poor in both urban and rural areas, it frequently endangers both people and the regional setting. The majority of those working in this industry either are uninformed of the risks, don't know about improved practices, or don't have access to investment capital to finance successful improvements.

3 Environmental Disturbances due to E-Waste

E-waste is both beneficial and damaging because of the huge amounts of precious materials it contains, as well as the fact that it is one of the primary sources of potential environmental concerns (Wäger et al. 2011). Concerns about improper e-waste recycling practices, particularly in developing nations, are growing as the global e-waste trade grows both legally and illegally. This increased environmental contamination threatens ecosystems and people who live in or close to key recycling sites. In China and India, two nations where informal e-waste recycling plays a significant economic role, levels of polybrominated dioxins and furans, polychlorinated dioxins and furans (PBDD/Fs and PCDD/Fs), polybrominated

Table 1 E-waste generation annually by different states of India (Singh and Gangeya 2020)

Top e-waste generator states in India	State-wise capacity recycling/dismantling (metric tons per annum as per CPCB)	Annual e-waste generation by states (in %)
Maharashtra	118,031.5	13.9
Tamil Nadu	130,636	9.1
Andhra Pradesh	44,002.5	8.7
Uttar Pradesh	4219.47	7.1
West Bengal	2640	6.9
Delhi	1989	6.7
Karnataka	126,015.48	6.2
Gujarat	128,604.92	6.1
Madhya Pradesh	13,600	5.3
All other states	–	30

diphenyl ethers (PBDEs), and lead (Pb) were measured in various environmental compartments. The risk to the people handling e-waste and the environment is increased by a lack of health and safety regulations as well as insufficient recycling techniques such as dumping, dismantling, incorrect shredding, burning, and acid leaching. Because some chemicals tend to long-distance transfer, the impact of pollutants produced by e-waste recycling operations is significant at the regional level. Despite the frightening facts, the situation can be quickly remedied by employing more responsible recycling methods as well as developing and enforcing e-waste-related national legislation, which includes prohibiting unregulated e-waste exports from developed countries (Chabhadiya et al. 2021). Despite numerous regulations, the informal sector's illicit activities, such as acid leaching, open incineration, and illegal dumping, have a significant impact on the environment, natural resources, and the safety of unorganized and unskilled labor. To develop a low-carbon, circular economy, there is a fundamental need for stakeholders to understand consumer behavior, global concerns, and possibilities in this field (Murthy and Ramakrishna 2022). In the last 15 years, the collection and recovery of e-waste have increased significantly throughout the world, yet comprehensive studies evaluating the environmental costs and benefits of these systems are still limited. One of the major environmental management problems that are growing as quickly as global consumption and population is waste. The laws and regulations governing the discharge of e-waste into the land, water, and air are getting stricter, which raises the cost of effective treatment and turns the dangerous substance into a usable good. Waste trading accounts for around 15% of the total trade in the European Union (Pathak et al. 2019). Authorities expressed significant concern about China, the world's largest CO_2 emitter, which accounts for roughly 29.4% of the total emissions, in the most recent Paris Agreement, when they agreed to cut global emissions to a specific level. As of 2019, coal accounted for more than 65% of China's total energy consumption (Abbasi et al. 2022).

4 Correlation Between Environment and Economics

The e-waste legislation has received support from about 71% of the world's population, but there is still a need to enforce and put in place a global legal framework. The formal sectors lack the infrastructure, equipment, and experience needed to collect and sustainably handle the growing amount of e-waste (Murthy and Ramakrishna 2022). The infrastructure, knowledge, and resources required by the official sectors to collect and properly manage the growing volume of e-waste are lacking. The first proposed public policy approach is a system that divides financial accountability among consumers, businesses, and the government. The system includes a deposit that is refunded to clients as an incentive for turning in e-waste. The second option is to include an e-waste conduit where it makes sense, bridging the official and informal sector divide. The informal sector would manage the collection, while the formal sector would handle recycling and disassembly. Due to the growing

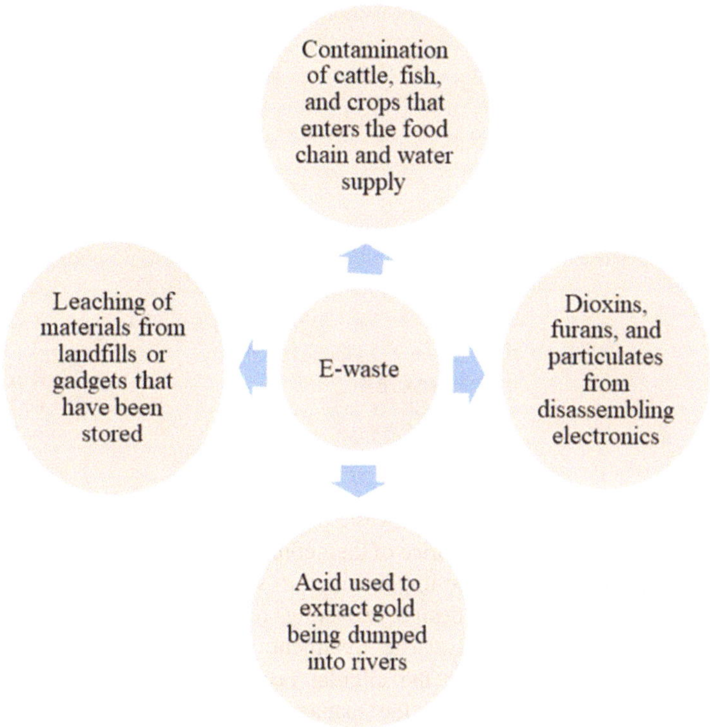

Fig. 2 Environmental imbalance by e-waste

demand for raw materials, notably rare earth elements, and in undeveloped and underdeveloped countries, there are unregulated e-waste recycling businesses. E-waste management challenges are becoming increasingly important. Construction of the infrastructure needed for electronic waste handling and credibility, as well as efficient material recovery processes and product design, is extremely challenging despite growing markets for recovered materials (Yu et al. 2010). With the growing demands for raw materials from the e-wastes and people's urge of using updated versions of electrical equipment and gadgets, environmental concerns like leaching of materials from landfills and contamination of water bodies, soil, and the air are also increasing consequently (refer Fig. 2).

Carbon emissions are strongly influenced by GDP; with a 1% increase in GDP, environmental change increases by 0.50% over time, according to statistical evidence. Pollution of the environment is frequently caused by economic practices in developed countries, as these economies rely on fossil fuels (oil, gas, coal, and nuclear energy) for mining and other commercial activities that raise CO_2 emissions in society as shown in Fig. 3 (Abbasi et al. 2022). E-waste statistics are crucial for more than simply environmental reasons; they also have a notable economic impact. In 2016, the projected total worth of all raw materials identified was around 55 billion euros from the generated e-waste which was higher compared to the global average

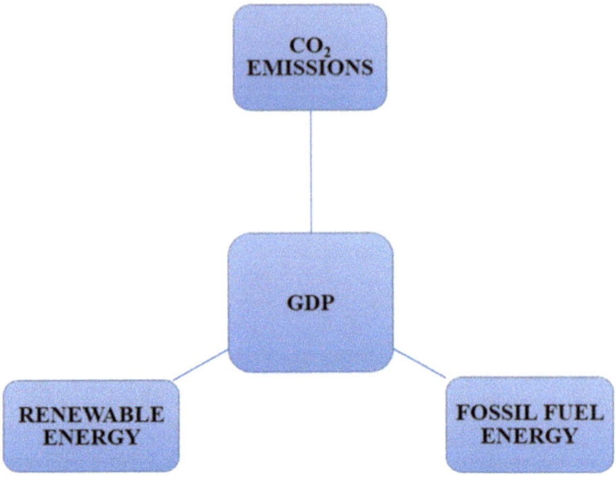

Fig. 3 Correlation of GDP with CO_2 emission and renewable energy

of the same year for GDP. Following waste management, the worth of recycled raw materials is much lower than the price of their components or old equipment. Models of the circular economy must be implemented to support closing the material loop through improved component design, recycling, reusing, and so on, as well as to reduce environmental harm. E-waste treatment has tremendous economic and employment potential owing to the circular economy concept. This entails the creation of strong legislation for the management of e-waste, which must be supported by data demonstrating the benefits to the environment and the economy.

4.1 Global Legislations and Initiatives of E-Waste

Legislations address the challenges associated with e-waste from multiple angles. The worldwide marketplace producing raw e-waste is anticipated to be worth $57.0 billion. Only 15.0 million tons (Mt) of CO_2 are offset by sustainable e-waste recycling and recovery, which is valued at $10.0 billion. The main hurdles to e-waste treatment include trash collection, sorting, inhomogeneity, low energy density, stopping the development of further waste, emissions, and cost-effective recycling. Only 78 countries have laws governing e-waste. In most cases, such legislation is not properly enforced. In poor countries such as Southeast Asia and Northern Africa, there is little to no e-waste legislation (Abbasi et al. 2022). Over 200 nonoperational gadgets were abandoned at various recycling facilities in the United States, and BAN (Basel Action Network) monitored them; few of the initiatives on e-waste management are shown in Table 2.

A 32.5% export rate was seen for the tracked equipment, and 31% was likely to be shipped illegally (Kumar et al. 2017). Laws have been enacted around the world to develop and execute effective and sustainable systems for collecting, recycling, and transferring e-waste (refer to Fig. 4). While the Restriction of Hazardous

Table 2 Different e-waste initiatives worldwide

Legislation/initiatives	Specification
Basel Convention and Basel Ban (1992)	Since 1992, a global agreement has been in effect governing the transfer of hazardous wastes, including e-waste, between nations
Canada's Electronics Product Stewardship (EPS)	EPS Canada was founded to work with organizations and the government to deliver a flexible, useful Canadian solution. This industry-led organization's initial members include 16 of the leading electronics manufacturers
Computer Take Back Campaign, Silicon Valley Toxics Coalition, and Basel Action Network	A collection of US-based nongovernmental organizations (NGOs) that work together on e-waste issues, such as worldwide advocacy for the Basel Ban, domestic collection and recycling initiatives, and in-depth research to promote local solutions for hazardous waste management
Initiative for the Stewardship of National Electronic Products (NEPSI)	A multi-stakeholder conversation to develop the framework for an American national e-waste management system. Participants in the NEPSI discussion include manufacturers, retailers, state and local governments, recyclers, environmental organizations, and others
European Recycling Platform (ERP) (2002)	The alliance was established by Hewlett Packard, Sony, Braun, and Electrolux at the end of 2002 to aid manufacturers in adhering to the e-waste guideline. One of its goals is to evaluate, set up, and operate a pan-European platform for waste and recycling management services
SECO/Empa E-Waste Scheme (2003)	A project to assess and improve e-waste recycling systems in various parts of the world through system analysis and knowledge exchange on recycling techniques and frameworks was established in 2003 by SECO (Swiss State Secretariat for Economic Affairs) and carried out by Empa (Swiss Federal Laboratories for Materials Testing and Research) in collaboration with several local partners and authorities
Solving the E-Waste Issue Through the StEP Program (2004)	The UN started an initiative to develop a global platform for nations to exchange ideas and establish e-waste systems to enhance and coordinate various international operations on the reverse supply chain at the (Electronic Goes Green) Conference in Berlin in 2004

Substance Directive (RoHS) regulation prohibited the use of particular toxic substances in the production of e-waste, the European e-waste directive was established in 2002 in the European Union to control end-of-life electronics to increase the collection and the efficiency of the recycling chain. The established collection goals

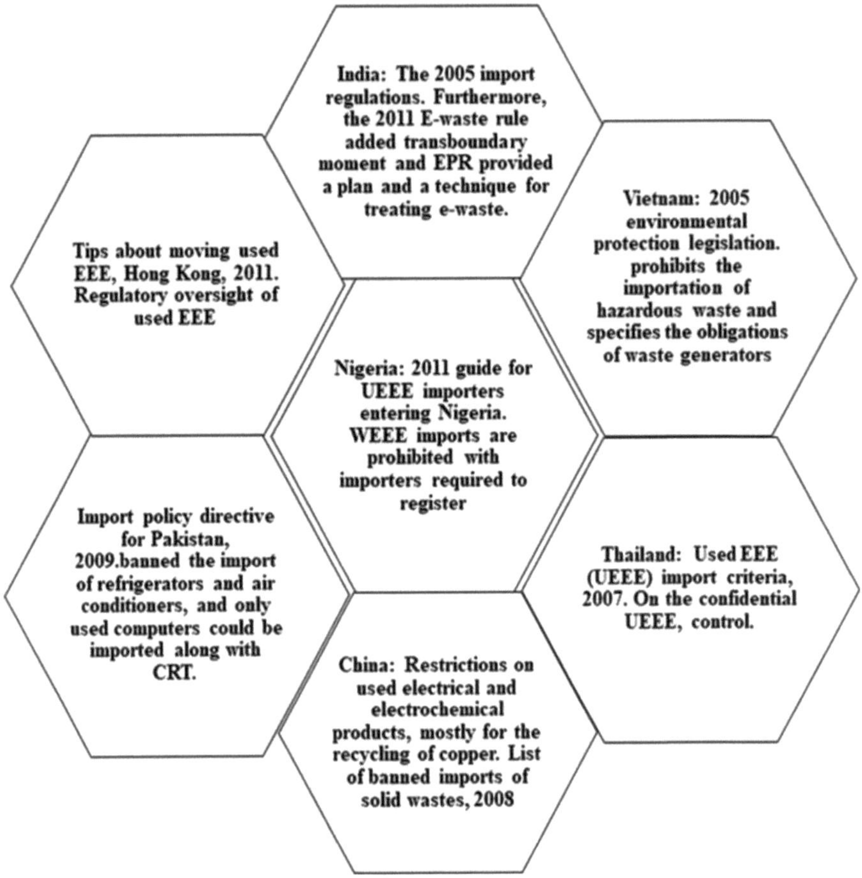

Fig. 4 India and other countries with their e-waste legislation

are expressed as a constant weight per resident (now 4 kg). The rules were modified in 2016, including the collection objective, which was set at 45% of the e-waste sold that year. Due to a paucity of land for adequate disposal of solid waste, in an effort to increase recycling rates, Japan passed the laws like the Small Appliance Recycling Law and the Household Appliance Recycling Law (HARL). More than 50% of all e-waste generated through Western countries is estimated to be illegally transferred to China. The Philippines, Malaysia, Nigeria, Ghana, Vietnam, India, Pakistan, and even Mexico and Brazil are also claimed to be receiving substantial amounts of e-waste (Golev et al. 2016; Vanegas et al. 2017). China introduced an extended producer responsibility policy for e-waste recycling in 2011. In order to collect and recycle electronic waste, standards for e-waste management and handling were developed in 2011. In Indonesia, e-waste management is not officially regulated by legislation, but it is managed by the Republic of Indonesia Act on Environmental Protection and Management as a Hazardous and Toxic Waste (Yoshida et al. 2016).

4.2 Indian Scenario

India established the guidelines for environmentally sound treatment of e-waste in 2011 to classify garbage based on its compositions and components. The 2016 amendment to the e-waste (management) regulations became law in India in 2018. India's e-waste policy prioritizes environmental concerns over economic concerns, opting for a single extended producer responsibility (EPR) approach over a diversified one as shown in Table 3.

To properly harness the urban mining potential of e-waste, existing restrictions must be made more compatible with successful techniques in other countries (Mudali et al. 2021). The electronics industry is the world's largest and most rapidly developing manufacturing sector. As a result, the management of e-waste has become a serious concern in the modern day. Some developed countries, such as Japan, have enacted complex legal procedures and regulations (such as the home appliance recycling law) to recover materials from e-waste, conserve resources, and minimize environmental damage. Global warming and climate change are currently being prioritized as significant environmental concerns on the policy agendas of both developed and developing countries. From this perspective, assessing e-waste recycling capability to a country's greenhouse gas (GHG) reduction goal would provide a new path for mitigating climate change. Two major gaps in current e-waste legislation are the lack of proper treatment for recovered materials and the absence of any regulations on materials to prevent heavy metals from entering new goods as shown in Table 4. To reduce the number of dangerous compounds entering downstream processes, it was advised that a knowledge base on the threat to the environment and ecotoxicology of these substances be illustrated and that advancements in the field of e-waste recycling get necessary. To address the growing volume of e-waste, manufacturers, recyclers, state and federal agencies, and the general public must work together.

The issue of bioaccumulation of heavy metals in water and soil and their entry into the food chain is brought on by the proliferation and potential toxicity of e-waste as shown in Table 4. The feasibility of resolving these challenges is a non-Organisation for Economic Co-operation and Development (OECD) scenario using EPR, an environmental policy principle that has been used for e-waste control in numerous OECD countries. Large gray markets for some electronic items and illegal e-waste imports are the two major impediments in the Indian context that could undermine EPR processes (Manomaivibool 2009).

A significant quantity of GHG emissions that would have been produced by the virgin production of materials might be avoided by implementing an adequate e-waste recycling and resource recovery scheme. For instance, recycling the unit weight of appliances such as televisions, air conditioners, refrigerators, and washing machines could help to reduce greenhouse gas emissions by 17.70, 27.34, 45.62, and 3.61 $kgCO_2eq$, respectively. The outcomes will be valuable for strengthening and implementing relevant legislation and regulations in nations throughout the

Table 3 Proposed strategies and future initiatives in India (Singh and Gangeya 2020)

Strategy 2023	Vision 2030	New initiatives	Future initiatives
Informal sector	Facilitate the integration of the informal sector into the supply chain for e-waste management while taking into account the workers' right to a living	Stringent provisions under extended producer responsibility	(Addressing the informal sector) Bridging the gap between formal and informal sectors
Policy instruments under EPR	Create a regularly updated, publicly accessible inventory of the amount of e-waste generated by the district, including the types of e-waste produced (such as computers, mobile phones, and appliances)	Cost-effective setting up recycling facilities	Access to environmentally sound technologies
Regulatory enforcement	To promote the widespread adoption of environmentally friendly e-waste recycling technologies, build a framework for policies governing the development of native technologies and/or technology transfer	Proper infrastructure	Creation of viable e-waste business models and execution of pilot initiatives for various advances
E-waste imports	Locate and capitalize on public policy tools that encourage manufacturers and producers to make investments in design for environmental adjustments to their product designs	Boosting the formal e-waste recycling industry	
Public awareness	Increase public knowledge of e-waste, its effects on society, the obligations placed on various stakeholders by current laws, and the righteous steps that individuals may take. These efforts should place more of an emphasis on prevention (i.e., lowering electronic device consumption) than on treatment (i.e., managing the generated e-waste)	Developing an online mass balance system: it will consist of all the key e-waste stakeholder's channelization: producers, importers, port authorities/customs, bulk consumers, PROs, dismantlers, and recyclers	

Table 4 Different pollutants from the e-waste polluting soil, air, and water

Environmental factors	Contamination of metals from e-waste	References
Air	Pb, Sn, Cr, As, Cd, benzo[a]pyrene, polycyclic aromatic hydrocarbons (PAH), polybrominated diphenyl ethers (PBDEs), polybrominated dibenzodioxins/furans (PBDD/Fs), polychlorinated dibenzodioxins/furans (PCDD/Fs), polybrominated dibenzodioxins /furans (PAH), flame retardant tetrabromobisphenol A (TBBPA)	Leung (2019), Needhidasan et al. (2014), Wu et al. (2016)
Soil	Pb, Ni, Zn, Cd, Mn, Hg, Cu, Cr, Sb, As, Al, PAH, polybrominated diphenyl ethers (PBDEs)	McGrath et al. (2017), Wu et al. (2015), Han et al. (2019), Arya et al. (2021)
Water	PAH, As, Cd, Ni, Cu, Pb, Zn, Mn	Yu et al. (2006), Wu et al. (2015)

Asia-Pacific and enhancing the methodical techniques of sound material recycling (Menikpura et al. 2014).

With a recycling rate of 16%, China produced more than 10 kt of electronic waste, making it the world's top producer. The United States came in second with a recycling rate of 15% and produced almost 7 kt as shown in Table 5. Globally, more than 54 Mt of electronic waste were produced in 2019; by 2050, that number is projected to double as much as 111 million tons yearly. Despite significant efforts to raise it, the average global collection rate for e-waste remains below average at about 20% (Parajuly et al. 2019).

5 Conclusions

Electronic waste is a global problem and became a serious environmental hazard. The world's fastest-growing waste stream is being produced by a rising population and an insatiable desire for modern items. It increases as a result of the economy's rapid expansion, people's strong need for a digitally linked society, green energy projects, and quick pace of technological progress, which promotes the obsolescence of goods. These wastes, in addition to being detrimental to the environment, result in the loss of rare and valuable materials like gold, silver, copper, platinum, palladium, as well as rare earths. Large amounts of electronic waste go unutilized since only 20% of it is managed appropriately globally. To tackle this problem, the Government of India came up with the policy of extended producer responsibility so that collection and recycling rates of the e-wastes can be improved. The volume of e-waste alone represents a substantial economic potential with a material worth of $62.5 billion. E-waste is therefore a valuable resource for urban mining. When compared to mining under the earth's crust, resource extraction from e-waste uses

Table 5 Top ten e-waste-generating countries with their GDP, recycling rate, types of waste, and volume (Statista 2022)

Countries	GDP (in billion USD)	E-waste produced (Kt)	Recycling rate (%)	Types of e-waste	References
China	18,100.04	10,129	16	Household appliances, computer, smartphones	Yang et al. (2021),
United States	25,464.48	6918	15	Computer, TV, mobile phones	Shittu et al. (2021)
India	3386.4	3230	1	Computers, mobile phones, TV	Kiran et al. (2021), Dasgupta et al. (2017)
Japan	4233.54	2569	22	Computer, TV, printer, camera	Xavier et al. 2021
Brazil	1924.13	2143	0	Mobile phones, computer, TV	Gollakota et al. (2020)
Russia	2215.29	1631	6	Cathode ray tubes, computers, mobile phones	Andeobu et al. (2021)
Indonesia	–	1618	Not available	Computer, mobile phones, TV	Batool et al. (2019)
Germany	4075.4	1607	52	Refrigerators, computers, TV	Patil and Ramakrishna (2020)
UK	3070.6	1598	57	Computers, TV, mobile	Ghimire and Ariya (2020)
France	2784.02	1362	56	Computers, TV, printers, scanners	Bonifazi et al. (2021)

a lot less energy and emits a lot less CO_2. Furthermore, it offers unrivaled opportunities to workers and international firms. India is the third largest producer of e-waste in the world, producing approximately two million tons annually, but only 1.% of it is recycled. Unorganized industries handle more than 95% of all generated e-waste. Due to improved awareness of e-waste management, many nations are enacting e-waste legislation.

References

Abbasi KR, Shahbaz M, Zhang J, Irfan M et al (2022) Analyze the environmental sustainability factors of China: the role of fossil fuel energy and renewable energy. Renew Energy 187:390–402. https://doi.org/10.1016/j.renene.2022.01.066

Andeobu L, Wibowo S, Grandhi S (2021) An assessment of e-waste generation and environmental management of selected countries in Africa, Europe and North America: a systematic review. Sci Total Environ 792:148078. https://doi.org/10.1016/j.scitotenv.2021.148078

Arya S, Kumar S (2020) E-waste in India at a glance: Current trends, regulations, challenges and management strategies. J Clean Prod 271:122707. https://doi.org/10.1016/j.jclepro.2020.122707

Arya S, Rautela R, Chavan D, Kumar S (2021) Evaluation of soil contamination due to crude E-waste recycling activities in the capital city of India. Process Saf Environ Prot 152:641–653. https://doi.org/10.1016/j.psep.2021.07.001

Baldé, C.P., Angelo, E.D., Luda, V., Deubzer, O., et al. (2022). Global transboundary e-waste flows monitor. https://ewastemonitor.info/wp-content/uploads/2022/06/Global-TBM_webversion_june_2_pages.pdf

Batool R, Sharif A, Islam T, Zaman K et al (2019) Green is clean: the role of ICT in resource management. Environ Sci Pollut Res 26:25341–25358. https://doi.org/10.1007/s11356-019-05748-0

Bonifazi G, Fiore L, Gasbarrone R, Hennebert P et al (2021) Detection of brominated plastics from e-waste by short-wave infrared spectroscopy. Recycling 6(3):54. https://doi.org/10.3390/recycling6030054

Chabhadiya K, Srivastava RR, Pathak P (2021) Two-step leaching process and kinetics for an eco-friendly recycling of critical metals from spent Li-ion batteries. J Environ Chem Eng 9:105232. https://doi.org/10.1016/j.jece.2021.105232

Das P, Gabriel JCP, Tay CY, Lee JM (2021) Value-added products from thermochemical treatments of contaminated e-waste plastics. Chemosphere 269:129409. https://doi.org/10.1016/j.chemosphere.2020.129409

Dasgupta D, Debsarkar A, Hazra T, Bala BK et al (2017) Scenario of future e-waste generation and recycle-reuse-landfill-based disposal pattern in India: a system dynamics approach. Environ Dev Sustain 19:1473–1487. https://doi.org/10.1007/s10668-016-9815-6

Forti V, Baldé CP et al. (2020) The Global E-waste Monitor 2020- Quantities, flows, and the circular economy potential. https://www.itu.int/en/ITU-D/Environment/Documents/Toolbox/GEM_2020_def.pdf

Ghimire H, Ariya PA (2020) E-wastes: bridging the knowledge gaps in global production budgets, composition, recycling and sustainability implications. Sustain Chem 1:154–182. https://doi.org/10.3390/suschem1020012

Golev A, Schmeda-Lopez DR, Smart SK, Corder GD et al (2016) Where next on e-waste in Australia? Waste Manag 58:348–358. https://doi.org/10.1016/j.wasman.2016.09.025

Gollakota ARK, Gautam S, Shu CM (2020) Inconsistencies of e-waste management in developing nations – facts and plausible solutions. J Environ Manage 261:110234. https://doi.org/10.1016/j.jenvman.2020.110234

Han Y, Tang Z, Sun J, Xing X et al (2019) Heavy metals in soil contaminated through e-waste processing activities in a recycling area: implications for risk management. Process Saf Environ Prot 125:189–196. https://doi.org/10.1016/j.psep.2019.03.020

Ikhlayel M (2018) An integrated approach to establish e-waste management systems for developing countries. J Clean Prod 170:119–130. https://doi.org/10.1016/j.jclepro.2017.09.137

Kiran M, Shanmugam PV, Mishra A, Mehendale A et al (2021) A multivariate discrete grey model for estimating the waste from mobile phones, televisions, and personal computers in India. J Clean Prod 293:126185. https://doi.org/10.1016/j.jclepro.2021.126185

Kumar A, Holuszko M, Espinosa DCR (2017) E-waste: an overview on generation, collection, legislation and recycling practices. Resour Conserv Recycl 122:32–42. https://doi.org/10.1016/j.resconrec.2017.01.018

Leung AOW (2019) Environmental contamination and health effects due to e-waste recycling. Electron Waste Manag Treat Technol 335–362. https://doi.org/10.1016/B978-0-12-816190-6.00015-7

Li W, Achal V (2020) Environmental and health impacts due to e-waste disposal in China—a review. Sci Total Environ 737:139745. https://doi.org/10.1016/j.scitotenv.2020.139745

Manomaivibool P (2009) Extended producer responsibility in a non-OECD context: the management of waste electrical and electronic equipment in India. Resour Conserv Recycl 53:136–144. https://doi.org/10.1016/j.resconrec.2008.10.003

McGrath TJ, Ball AS, Clarke BO (2017) Critical review of soil contamination by polybrominated diphenyl ethers (PBDEs) and novel brominated flame retardants (NBFRs); concentrations, sources and congener profiles. Environ Pollut 230:741–757. https://doi.org/10.1016/j.envpol.2017.07.009

Menikpura SNM, Santo A, Hotta Y (2014) Assessing the climate co-benefits from waste electrical and electronic equipment (WEEE) recycling in Japan. J Clean Prod 74:183–190. https://doi.org/10.1016/j.jclepro.2014.03.040

Mudali UK, Patil M, Saravanabhavan R, Saraswat VK (2021) Review on e-waste recycling: part I—a prospective urban mining opportunity and challenges. Trans Indian Natl Acad Eng 6:547–568. https://doi.org/10.1007/s41403-021-00216-z

Murthy V, Ramakrishna S (2022) A review on global e-waste management: urban mining towards a sustainable future and circular economy. Sustainability 14:647. https://doi.org/10.3390/su14020647

Needhidasan S, Samuel M, Chidambaram R (2014) Electronic waste—an emerging threat to the environment of urban India. J Environ Health Sci Eng 12:1. https://doi.org/10.1186/2052-336X-12-36

Parajuly K, Kuehr R, Awasthi AK, Fitzpatrick C et al (2019) Future e-waste scenarios. https://collections.unu.edu/eserv/UNU:7440/FUTURE_E-WASTE_SCENARIOS_UNU_190829_low_screen.pdf

Pathak P, Pandey N (2023) Hydrometallurgical recycling of critical metals from spent Ni-Cd batteries with emphasis on the separation of Cd^{2+} over Ni^{2+} using D2EHPA. Geosystem Eng. https://doi.org/10.1080/12269328.2023.2201290

Pathak P, Srivastava RR, Ojasvi (2017) Assessment of legislation and practices for the sustainable management of waste electrical and electronic equipment in India. Renew Sustain Energy Rev 78:220–232. https://doi.org/10.1016/j.rser.2017.04.062

Pathak P, Srivastava RR, Ojasvi (2019) Environmental management of e-waste. Electron Waste Manag Treat Technol:103–132. https://doi.org/10.1016/B978-0-12-816190-6.00005-4

Pathak P, Sharma S, Ramkrishna S (2023) Circular transformation in plastic management lessen the carbon footprint of the plastic industry. Mater Today Sustain 22:100365. https://doi.org/10.1016/j.mtsust.2023.100365

Patil RA, Ramakrishna S (2020) A comprehensive analysis of e-waste legislation worldwide. Environ Sci Pollut Res 27:14412–14431. https://doi.org/10.1007/s11356-020-07992-1

Shittu OS, Williams ID, Shaw PJ (2021) Global E-waste management: can WEEE make a difference? A review of e-waste trends, legislation, contemporary issues and future challenges. Waste Manag 120:549–563. https://doi.org/10.1016/j.wasman.2020.10.016

Singh S, Gangeya (2020) E-waste Roadmap 2023 for India. https://greene.gov.in/wp-content/uploads/2020/12/2020120916.pdf

Srivastava RR, Pathak P (2019) Policy issues for efficient management of e-waste in developing countries. In: Handbook of electronic waste management: international best practices and case studies. Elsevier, pp 81–99. https://doi.org/10.1016/B978-0-12-817030-4.00002-4

Statista (2022) The 20 countries with the largest gross domestic product (GDP) in 2022 (in billion U.S. dollars). https://www.statista.com/statistics/268173/countries-with-the-largest-gross-domestic-product-gdp/

Tabelin CB, Park I, Phengsaart T, Jeon S et al (2021) Copper and critical metals production from porphyry ores and e-wastes: a review of resource availability, processing/recycling challenges, socio-environmental aspects, and sustainability issues. Resour Conserv Recycl 170:105610. https://doi.org/10.1016/j.resconrec.2021.105610

Vanegas P, Peeters JR, Cattrysse D, Dewulf et al (2017) Improvement potential of today's WEEE recycling performance: the case of LCD TVs in Belgium. Front Environ Sci Eng 11:13. https://doi.org/10.1007/s11783-017-1000-0

Wäger PA, Hischier R, Eugster M (2011) Environmental impacts of the Swiss collection and recovery systems for waste electrical and electronic equipment (WEEE): a follow-up. Sci Total Environ 409(10):1746–1756. https://doi.org/10.1016/j.scitotenv.2011.01.050

Wu Q, Leung JYS, Geng X, Chen S et al (2015) Heavy metal contamination of soil and water in the vicinity of an abandoned e-waste recycling site: Implications for dissemination of heavy metals. Sci Total Environ 506–507:217–225. https://doi.org/10.1016/j.scitotenv.2014.10.121

Wu Y, Li Y, Kang D, Wang J et al (2016) Tetrabromobisphenol A and heavy metal exposure via dust ingestion in an e-waste recycling region in Southeast China. Sci Total Environ 541:356–364. https://doi.org/10.1016/j.scitotenv.2015.09.038

Xavier LH, Ottoni M, Lepawsky J (2021) Circular economy and e-waste management in the Americas: Brazilian and Canadian frameworks. J Clean Prod 297:126570. https://doi.org/10.1016/j.jclepro.2021.126570

Yang WD, Sun Q, Ni HG (2021) Cost-benefit analysis of metal recovery from e-waste: Implications for international policy. Waste Manag 123:42–47. https://doi.org/10.1016/j.wasman.2021.01.023

Yoshida A, Terazono A, Ballesteros FC, Nguyen DQ et al (2016) E-waste recycling processes in Indonesia, The Philippines, and Vietnam: a case study of cathode ray tube TVs and monitors. Resour Conserv Recycl 106:48–58. https://doi.org/10.1016/j.resconrec.2015.10.020

Yu XZ, Gao Y, Wu SC, Zhang HB, Cheung et al (2006) Distribution of polycyclic aromatic hydrocarbons in soils at Guiyu area of China, affected by recycling of electronic waste using primitive technologies. Chemosphere 65(9):1500–1509. https://doi.org/10.1016/j.chemosphere.2006.04.006

Yu J, Williams E, Ju M, Shao C (2010) Managing e-waste in China: policies, pilot projects and alternative approaches. Resour Conserv Recycl 54(11):991–999. https://doi.org/10.1016/j.resconrec.2010.02.006

Zhang Z, Malik MZ, Khan A, Ali N et al (2022) Environmental impacts of hazardous waste, and management strategies to reconcile circular economy and eco-sustainability. Sci Total Environ 807:150856. https://doi.org/10.1016/j.scitotenv.2021.150856

Hazards Associated with Industrial Effluents and Its Mitigation Strategies

Ziaul Haque Ansari and Uttam Bista

Abstract Industrial effluent is related to liquid waste generated by industries that may be emitted into the municipal drainage or sewer system. The effluents composition is extremely variable and heavily influenced by the various industries from which they originate. Contaminants can be grouped into different classes such as endocrine disrupting compounds (EDCs), pharmaceuticals, pesticides, heavy metals and metalloids, per- and polyfluoroalkyl substances (PFAS), and microplastics. Each contaminant upon exposure possesses a specific health impact on humans and animals as well as on marine life when mixed in the sewer.

This harmful effluent needs to be treated to reduce its adverse effect either on-site or off-site. There are a variety of conventional and advanced wastewater treatment facilities available for use. For on-site installation of treatment facilities, the cost of such plants can be reduced by the government by taxing less on equipment purchases. For off-site installation of the treatment plant, the local body can seek financial support from the industries producing these effluents.

Keywords Effluent · Wastewater · Endocrine · Heavy metals · Per- and polyfluoroalkyl · Pharmaceuticals · Pesticides · Microplastics

Abbreviations

AMR	Antimicrobial resistance
BBP	Benzyl butyl phthalate
BPA	Bisphenol A
DBP	Di-n-butyl phthalate
DEHP	Di(2-ethylhexyl) phthalate
DiBP	Di-iso butylphthalate

Z. H. Ansari (✉) · U. Bista
Department of Applied Sciences and Chemical Engineering, Pulchowk Campus, Institute of Engineering (IoE), Tribhuvan University, Kathmandu, Nepal
e-mail: ziaul.ansari@pcampus.edu.np

P. Pathak et al. (eds.), *Anthropogenic Environmental Hazards*,
https://doi.org/10.1007/978-3-031-41013-0_5

DVFA	Danish Veterinary and Food Administration
EC	European commission
EDCs	Endocrine-disrupting compounds
EFSA	European Food Safety Authority
EPA	Environmental Protection Agency
kg bw	Kilograms of body weight
LOAEL	Lowest observed adverse effect level
MRL	Minimum risk level
PCBs	Polychlorinated biphenyls
PCDDs	Polychlorinated dibenzo-p-dioxins
PCDFs	Polychlorinated dibenzofurans
PFAS	Per- and polyfluoroalkyl substances
PFOA	Perfluorooctanoic acid
PFOS	Perfluorooctane sulfonate
POPs	Persistent organic pollutants
PTMI	Provisional tolerable monthly intake
RfD	Reference dose
TCDD	2,3,7,8- tetrachlorodibenzo-p-dioxin
TDI	Tolerable daily intake
TMI	Tolerable monthly intake
TWI	Tolerable weekly intake
WHO	World Health Organization
μg/kg	Microgram per kilogram

1 Introduction

Industries, a key factor of economic growth, also have caused serious pollution and environmental issues (Wen 2009; Wei and Huang 2001). Industrial waste refers to unwanted residual materials generated by industrial processes or operations. Waste materials can be in solid, liquid, or gaseous form (Wen 2009; Misra and Pandey 2005; Ojoawo et al. 2011) that include among others food wastes, packaging materials, ashes, smoke, rubbish, debris, special wastes, and hazardous wastes (Aivalioti et al. 2014; Abduli 1996; Tchobanoglous et al. 1993; Casares et al. 2005; Vigneswaran et al. 1999). Special wastes including clinical and pharmaceutical wastes are considered nonhazardous waste but possess unique regulatory requirements.

Wastes can be categorized as hazardous and nonhazardous according to their effect on human or other organisms. Hazardous materials as listed in the Resource Conservation and Recovery Act (RCRA) regulations can be toxic, flammable, ignitable, reactive, or corrosive. Oil, printing ink, paint, varnish, soluble cutting emulsion, and disinfectants are some examples of hazardous wastes. Wastes that do fall under the category of hazardous wastes as the Environmental Protection Agency

(EPA)'s definition are termed as nonhazardous waste. Laboratory waste consists of empty aerosol cans, nonsurgical and nonradioactive medical refuse, as well as food and packaging waste, which are a few examples of nonhazardous wastes.

In industrialized countries, public pressure regarding pollution led local bodies to act strictly, while in developing countries, awareness of pollution is lower in public, and the action of the local body is also not significant. Rapid industrialization and the use of toxic materials in processing have led to problems with environmental pollution (Wen 2009).

1.1 Solid Wastes

Annually, about 12 billion tons of industrial waste is being generated which will exceed upto 19 billion tons soon (Li 2009; Pappu et al. 2007; Yoshizawa et al. 2004). They can be of hazardous and nonhazardous nature (Li 2009). Unwanted materials produced during processing depend upon the types of industries. Mining industries produce waste stones, metallurgical industries produce slag, power industries produce ash, and chemical industries produce inferior products, unreacted materials, and disabled catalysts. Oil chemical industries produce oil mud and slag.

1.2 Liquid Wastes

Industrial liquid wastes include feedstock materials, by-products, product material in soluble or particulate form, washing and cleaning agents, solvents, etc. These wastes may be nontoxic inorganic substances or toxic organic substances. The effluents or wastewater including these materials pass through sewer network and affect the aquatic environment (Fig. 1).

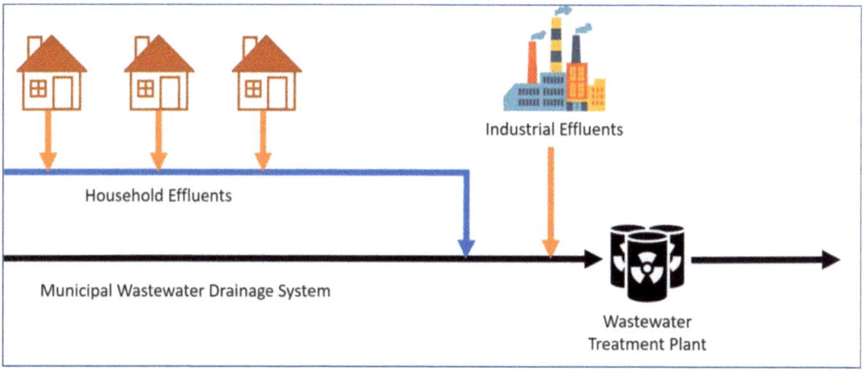

Fig. 1 Schematic diagram of domestic and industrial effluent

1.3 Gaseous Wastes

Industrial emissions are in the form of particulate vapor, powder, mineral fumes, and smoke. Depending on the local regulations, they can be emitted into the atmosphere with or without prior treatment. The components in the emissions depend upon the types of fuel used and the nature of industries (Ojoawo et al. 2011). The environmental impacts of waste, such as pollution and disease, have turned into a source of major concern. When inadequately managed, wastes may lead to the deterioration of sewage systems with increasing environmental pollution and diseases (Hasheela 2009).

Sewage effluent-containing pathogens led to diseases such as diarrhea, polio, meningitis, and hepatitis. WHO (2003) estimated in 1998 that approximately 1.8 million children passed away in developing nations due to microorganisms, the vast majority of which originated from contaminated food and water. It is estimated that 12 million people die annually around the globe due to improper waste management (Davidson et al. 1992). Humans come into contact with wastes or are exposed to wastes by different routes such as ingestion, inhalation, and absorption. Drinking water and food containing hazardous substances from refuse residues constitute an ingestion route, the inhaling route is breathing airborne wastes, and the absorption route is by direct contact with waste residues (Hinga and Batchellor 2005).

Hazardous industrial wastes may have short- or long-term effects on humans and ecological systems. The potential health effects on human depend upon the characteristics of the hazardous chemicals, the duration of exposure, the health status of the exposed individual, as well as weather conditions (Misra and Pandey 2005; Grisham 1986). Health effects on human due to exposure to hazardous wastes may be carcinogenesis (i.e., causing cancers), genetic defects, reproductive abnormalities, alterations of immunobiological homeostasis, central nervous system (CNS disorder), and congenital anomalies (El Sidig NOA 2004). To prevent the adverse effect of hazardous materials or wastes on human and the environment, proper storage and treatment and disposal by the latest technology are required.

At present, waste management is one of the world's greatest environmental challenges (Kan 2009). Industrial wastes can be toxic, ignitable, corrosive, and reactive substances that need to reduce the adverse effect before being discharged into the environment (Zurbrugg 2002). Waste management methods differ from developed countries to developing countries and also differ in urban to rural areas (Addo 2013). Waste is managed in accordance with public health, economics, engineering, conservation, esthetics, and other environmental considerations and public attitudes (Tchobanoglous et al. 1977; Demirbas 2011).

Solid waste management covers control of the generation, storage, collection, transfer and transport, processing, and disposal of solids. Liquid waste management deals with wastewater treatment and sewage treatment. A variety of waste streams is generated by industries, referred as industrial effluents, which are discharged to either the municipal or public effluent treatment system or directly to receiving waters. These effluents contain numerous substances that may pose health hazards to humans. Depending on the industries, to reduce contaminant concentrations effluents may be treated on-site, or are discharged to municipal effluent treatment system or directly to receiving waters. Countries possess policies regarding discharging effluents at tolerable levels and determining concentration limits for a variety of potentially hazardous contaminants. Additionally, sewage sludge, a potentially hazardous substance that settles out of domestic and industrial wastewater during remediation, also poses a threat to public health (Eaton 2022).

Different countries have set their own standards for concentration limits for several potentially harmful contaminants present in the discharge effluent. Also, the land is polluted by the addition of unstabilized sewage sludge from the treatment plant. The finding of new contaminants, their harmful effects, and their discharge limits will keep on being researched. Using global data, this report attempts to determine contaminants that are of more concern to human health and the types of industries that are discharging them. Depending on international data, contaminants that are of most concerns for human health are reported here (Stewart et al. 2016). The considered contaminants fall into seven general categories.: endocrine-disrupting compounds (EDCs), heavy metals and metalloids, per- and polyfluoroalkyl substances (PFAS), pharmaceuticals, pesticides, and microplastics. The potential health concern associated with each contaminant is discussed. However, the precise health hazard they pose is currently unclear and uncertain.

2 Endocrine-Disrupting Compounds

These are substances that interfere with biosynthesis, metabolism, or action of hormones to disrupt normal hormone signaling. (Diamanti-Kandarakis et al. 2009). The above chemicals of various categories have associated with variety of health issues and are considered a public health concern (Zoeller et al. 2014). These compounds are found in food, consumer goods, and pharmaceuticals (such as birth control medication), which upon discarded by human through excretion or activity pass to the municipal wastewater.

2.1 Nonylphenol and Nonylphenol Ethoxylates

Nonylphenol is a synthetic alkylphenol used in the production of antioxidants and lubricating oil additives and nonylphenol ethoxylate surfactants (Soares et al. 2008). Such surfactants are extensively used in cosmetics and coatings, detergents and cleaning products, degreasers, emulsifiers, and wetting and de-wetting agents (Environment Canada and Health Canada 2001; Soares et al. 2008). On account of their widespread use, considerable amounts of NPEs enter into both industrial and residential wastewater systems. Nonylphenol and its ethoxylates have high environmental persistence. The estimated half-life of NPEs in sediments is greater than 60 years (Shang et al. 1999), they bioaccumulate over aquatic creature (Gautam et al. 2015), and also been identified in human lactation (Sise and Uguz 2017).

2.1.1 Health Hazard

The studies of laboratory animals reflect that nonylphenol can have a negative impact on reproduction and the immune and nervous systems (Cressey 2018). Danish Veterinary and Food Administration (DVFA) (Nielsen et al. 2000) and European Commission (EC) (2002) have recommended nonylphenol and its ethoxylates exposure limits which is represented in Table 1.

2.1.2 Sources of Industrial Effluent

NPEs is mostly used for processes such as wool scouring, bleaching, laundering, and dyeing in Textile industries (Ho and Watanabe 2017). The concentrations range of 0.23–26 μg/L of nonylphenol was found to be in textile industry effluent which was discharged to the municipal wastewater management system of Canada (Environment Canada and Health Canada 2001). NPEs are used in paper and pulp industry for wetting of pulp fibers, and in the leather industry (Groshart et al. 2001).

2.1.3 Limit of Discharge and Policies

In the European Union (EU), the discharge limit under Directive 2008/105/EC in surface water is 2 μg/L for nonylphenols. Canadian Environmental Protection Act (1999) has classified both nonylphenol and its ethoxylates as harmful compounds under Schedule 1. In the United States, the Environmental Protection Agency (EPA) set the rule of Agency review for its use. In New Zealand, 50 mg/kg dry weight is the proposed concentration limit of nonylphenol and its ethoxylates (Water New Zealand 2017).

2.2 *Bisphenol A*

It is a synthetic chemical which is mainly used for the manufacture of epoxy resins, polycarbonate plastics (NIEHS 2021), as well as food storage vessels (Cressey 2018).

Table 1 Recommended nonylphenol and its ethoxylates exposure limits

Substance	LOAEL EC (mg/kg bw/day)	TDI DVFA (mg/kg bw/day)	LOAEL DVFA (mg/kg bw/day)
4-Nonylphenol (NP)	15	0.005	15
Nonylphenol ethoxylates (NPEs)	–	0.013	40

2.2.1 Health Hazard

New Zealand Environmental Protection Authority suspected Bisphenol A could harm fertility or a fetus (EPA 2021). United States Food and Drug Administration (EPA 2021) and European Food Safety Authority (EFSA 2015) diminished the tolerable daily intake (TDI) of BPA to 5 and 4 μg/kg body weight/day respectively. However, a recent study reflects effects on concentrations as minimum as 2.5 microgram per kilogram of body weight per day (Heindel et al. 2020).

2.2.2 Sources of Industrial Effluent

Municipal wastewater contains BPA that comes from food and beverage packaging. Industrial effluent from paper mills (Balabanic and Klemencic 2011; Fuerhacker 2003; Lee and Peart 2000; Lee et al. 2015), textile industries (Lee and Peart 2000; Pothitou and Voutsa 2008), tanning industries (Pothitou and Voutsa 2008), metal or wood industries, chemical industries, dry cleaning/cloth washing, plastics and polymer industries (Fuerhacker 2003; Lee and Peart 2000), and petrochemical industries (Mirzaee et al. 2019) contains subsequent amount of BPA.

2.2.3 Discharge Limits and Regulation

The set limit for BPA in industrial effluent is 1.75 μg/L in Canada (Government of Canada 2018), whereas in the United States, BPA action plan has been created by EPA (US EPA 2021a, b, c, d, e, f, g, h) without a discharge concentration limit.

2.3 Phthalates

Phthalates or phthalate plasticizers are used for solvent properties to make products more durable and flexible. They are used in various products such as vinyl flooring, plastic packaging, medical tubing, shampoos, hair sprays, soaps (CDC 2021), and cosmetics (U.S. Food and Drug 2021). Phthalates, which have a different types of derivatives, are diesters of 1,2-benzenedicarboxylic acid. Four derivatives included in discussion are BBP, DBP, DiBP, and DEHP.

2.3.1 Health Hazards

Phthalate syndrome is a common health effect that refers to the capacity to inhibit androgen biosynthesis, thereby disrupting sexual differentiation of male (CHAP 2014; National Research Council 2008). The health hazard posed by above listed

Table 2 Phthalate health implications and intake limits

Phthalate compound	RfD (mg/kg bw/day)	RPF (EFSA 2022a, b)
BBP	0.2 (US EPA 1989)	0.1
DBP	0.1 (US EPA 1987a, b)	5
DEHP	0.02 (US EPA 1987a, b)	1
DiBP	–	–

phthalates is reproductive or developmental toxicity (antiandrogenic) concerns (Ashworth and Chappell 2015). Other report suspected that they damaged fertility or the unborn child (Cressey 2018). The TDI for phthalates is set as 0.05 mg/kg bw/day by EFSA by setting the index compound as DEHP and expressing potency of another phthalate's relative to DEHP (EFSA 2022a, b) (Table 2).

2.3.2 Sources of Industrial Effluent

In Slovenia, DBP, BB, and DEHP were found in the waste from paper and pulp industries (Balabanic and Klemencic 2011). In Argentina and India, BBP, DBP, DEHP, and DiBP were found in the effluent from tanneries (Bharagava et al. 2018; Zubair Alam et al. 2010; Labunska et al. 2011). France found the DBP, BBP, and DEHP in the waste generated from the textile industries, pharmaceutical industries, aerospace company, waste management, vehicle washing, cosmetics products, metallurgy, and transportation maintenance industries (Bergé et al. 2014). Effluents from a turkey processing plant also reported DiBP concentration (Buyukada 2019).

2.3.3 Limit of Discharge and Policies

There are regulations regarding phthalate levels in kids toys, cosmetics kids care products etc., information about its level in sewage is not sufficient (Government of Canada 2017). In United States, EPA set discharge limits for DEHP, BBP, and DBP summarized in Table 3 (US EPA ELGs 2022a, b, c).

2.4 Dioxins

Dioxin, or 2,3,7,8- tetrachlorodibenzo-p-dioxin (TCDD), refers to chemicals polychlorinated dibenzofurans (PCDFs), polychlorinated dibenzo-p-dioxins (PCDDs), and polychlorinated biphenyls (PCBs). Stockholm Convention concluded these three classes of dioxin as persistent organic pollutants (POPs). It means signatories must act either to reduce or minimize or eliminate its release where possible. Chlorine-containing industrial processes produces dioxin as unwanted

Table 3 Description of phthalate compounds US EPA effluent limits

Phthalate compound	Source	Max concentration daily (μg/L)	Max monthly average (μg/L)
DEHP	Plastics, organic substances, and synthetic fibers	258–279	95–103
	Centralized waste management	215–267	101–158
DBP	Plastics, organic substances, and synthetic fibers	43–57	20–27
BBP	Centralized waste management	188	BBP

Table 4 Recommended dioxin and dioxin-like compound exposure limits

RfD (US EPA 2021) (ng/kg bw/day)	TWI (EFSA 2018) (ng/kg bw/week)	PTMI JECFA (WHO 2019) (ng/kg bw/month)	TMI New Zealand (MoH 2020) (ng/kg bw/month)
0.0007 (for TCDD)	0.002	0.07	0.03

by-products from pulp and paper industries, herbicide/pesticide production, and smelting.

2.4.1 Health Hazards

There is sufficient evidence that dioxin causes among others chloracne, non-Hodgkin and Hodgkin disease, hypertension, etc. (Ministry of Health 2020). Table 4 represents the recommended dioxins and dioxin-like compounds exposure limits set by different agencies, such as US EPA, Joint Food and Agriculture Organization of the United Nations (FAO)/WHO Expert Committee on Food Additives (JECFA), EFSA, and New Zealand Ministry of Health.

2.4.2 Sources of Industrial Effluent

Effluent and sludge from pulp and paper mills contain dioxins as they perform chlorine bleaching (Whittemore et al. 1990). Dioxin is also found in the effluents from other manufacturing industries such as caprolactam (an intermediate of nylon), vinyl chloride, acetylene, alumina fibers, chlorobenzene, 4-chloro sodium hydrogen phthalate, 2,3-dichloro-1,4-naphthoquinone, and organic colored pigments (Kawamoto and Weber 2021). Dioxins were found in the effluents of chemical industries that manufacture petroleum, plastics, synthetic polymers, nonvolatile elastics, industrial organic chemicals, cyclic organic crudes, dyes and pigments, pesticides, and agricultural chemicals (Sappington et al. 2015).

2.4.3 Limit of Discharge and Policies

In Japan, the PCDD/PCDFs discharge limit is set to 10 picogram toxic equivalentper litrefor industrial effluent (Kawamoto and Weber 2021). Canadian Environmental Protection Act, 1999 (Government of Canada 2021) prohibited pulp and paper industry from releasing 2,3,7,8-TCDD and 2,3,7,8-TCDF into the environment. For 2,3,7,8-TCDD and 2,3,7,8-TCDF by the pulp, paper and paperboard industries EPA has set discharge limit of maximum 10 pg/L per day (US EPA ELGs 2022a, b, c). Trade Waste Standard (Standards New Zealand 2004) set a maximum concentration of 0.002 g/m^3 (2 μg/L) for PCBs.

3 Heavy Metals and Metalloids

Heavy metals are high density (i.e., minimum five times denser in comparison with water) metallic elements (Tchounwou et al. 2012). Industrial effluents may contain chromium, nickel, lead, copper, metalloid arsenic, zinc, and cadmium (Wang 2018) and in sewage sludge as bio-solids.

3.1 Cadmium

It is a harmful and undesirable heavy metal present in phosphate rock, ingredient of superphosphate fertilizer. Cadmium enters the agricultural land through the use of fertilizer and industrial effluents containing cadmium to wastewater.

3.1.1 Health Hazards

Despite lowest levels, cadmium is toxic to body. It damages kidneys and gets accumulated with a half-life of approximately 15 years (Mannetje et al. 2018). Cadmium is used in various industries such as textile, electronics, electroplating, chemical, metal finishing, and metallurgical industries (Velusamy et al. 2021).

Table 5 represent the recommended exposure limits set by different agencies, such as Dutch National Institute of Public Health and the Environment (RIVM), JECFA, US EPA, European Food Safety Authority (EFSA), and Food Standards Australia New Zealand (FSANZ), US Agency for Toxic Substances and Disease Registry (ATSDR).

Table 5 Recommended cadmium exposure limits

TDI RIVM (Baars et al. 2001) (ng/kg bw/day)	RfD (EPA 2021) (ng/kg bw/day)	PTMI JECFA (WHO 2021a, b) (ng/kg bw/month)	TWI (EFSA 2021) (ng/kg bw/week)	PTMI (FSANZ 2011) (ng/kg bw/month)	MRL (ATSDR 2021) (ng/kg bw/day)
500 (oral)	500 (water) 1000 (food)	25,000	2500	25,000	100 (chronic oral)

Table 6 US EPA cadmium discharge limits summary

Point source	Daily maximum concentration (mg/L)	Average concentration [monthly] (μg/L)	Max average concentration [monthly] (μg/L)
Electroplating	1.2	–	–
Inorganic chemicals production	0.84	0.28	–
Metal coating	0.11–0.69	0.07–0.26	–
Centralized waste management	0.0172–0.782	–	0.0102–0.163
Ore mining and dressing	0.1	0.05	–
Electronic components	0.06–0.55	–	–

3.1.2 Limit of Discharge and Policies

The summary of discharge limits set by EPA (US EPA ELGs 2021) is represented in Table 6.

3.2 Chromium

Chromium (Cr) is found in oxidation states such as most stable chromium VI and chromium III (Wilbur et al. 2012). An essential nutrient chromium III is found naturally, whereas highly toxic chromium VI seldom occurs naturally which readily occur reduction reaction to chromium III (US EPA 1984). Anthropogenic activities produced chromium VI which when enters water form relatively stable (US EPA 1984; Wilbur et al. 2012).

Table 7 Recommended chromium III and VI exposures limits

	TDI RIVM (Baars et al. 2001) (mg/kg bw/day)	RfD (mg/kg bw/day)	TDI CONTAM (EFSA 2014a, b) (mg/kg bw/day)	TDI (WHO IPCS 2013) (mg/kg bw/day)	MRL (ATSDR 2021) (mg/kg bw/day)
Cr III	0.005 (water soluble) 5 (insoluble)	1.5 (EPA 2021) (insoluble)	0.3	–	–
Cr VI	0.005* (oral)	0.003 (EPA 2021) (oral)	–	0.0009 (oral)	0.0009 (chronic oral)

*Provisional Maximum Permissible Risk, noncarcinogenic effects

3.2.1 Health Hazards

The exposure of chromium VI may affect on the respiratory system and kidneys and also causes cancer (Mannetje et al. 2018). Different organizations have set recommended exposure limits which are mentioned in Table 7.

3.2.2 Sources of Industrial Effluent

Chromium is present in the effluent of industries such as textiles, metal finishing and electroplating, tanneries, dyes and pigment, wood preservation, and fertilizer industries. (Dermentzis et al. 2011, Verma et al. 2013). Very high level of chromium in the effluent is reported from the electroplating industry (reportedly up to 2500 mg/L of the highly toxic chromium VI) (Dermentzis et al. 2011), substantial chromium concentrations in textile dying wastewaters (Çetin et al. 2008), and leather tanneries (0.2 to more than 14 mg/L) in Argentina (Labunska et al. 2011).

3.2.3 Limit of Discharge and Policies

In the European Union, discharge limits of total chromium vary among member states, with a maximum discharge limit of 5 mg/L for total chromium and 1 mg/L for chromium VI in water (Vaiopoulou and Gikas 2020).

In the US, discharge limit is set for total chromium on daily maximum concentration for electroplating 7 mg/L, leather tanning and finishing (12–19) mg/L, and timber 4 mg/L (US EPA ELGs 2021). The discharge limits for total chromium are summarized in Table 8.

Table 8 US EPA total chromium discharge limits summary

Point source	Daily maximum concentration (μg/L)	Average concentration [monthly] (μg/L)	Max average concentration [monthly] (μg/L)
Electroplating	7000	–	–
Inorganic chemicals production	230–3000	120–1200	–
Metal coating	2770	1710	–
Centralized waste management	167–15,500	–	52.2–3070
Electronic components	560–650	–	260–300

3.3 Lead

Lead (Pb) in environment can be found in trace amounts. In early days, industries used it extensively in products such as ceramics, cosmetics, petrol, paints, batteries, and plumbing materials (EPA 2021). Its uses have been phased out because of its toxicity.(Pickston et al. 1985).

3.3.1 Health Hazards

Lead (Pb) exposure is associated with a several adverse health hazard, which include increased blood pressure, decrease in renal function and fertility, and neurocognitive effects. Neurodevelopmental effects in children even at low level (Mannetje et al. 2018). The exposure limit reported by RIVM is 3.6 μg/kg bw/day, oral (Baars et al. 2001).

3.3.2 Sources of Industrial Effluent

Lead in the effluent from the iron, steel, pulp and paper industries (US Environmental Protection Agency 2018), paint industries (Malakootian et al. 2009), a brewery, and a textile industry (Muhammd et al. 2018) are found.

3.3.3 Limit of Discharge and Policies

The discharge limits set by US EPA ELGs are summarized in Table 9.

Table 9 US EPA lead discharge limits summary

Point source	Daily maximum concentration (μg/L)	Average concentration [monthly] (μg/L)	Max average concentration [monthly] (μg/L)
Electroplating	600	–	–
Inorganic chemicals production	180–3400	48–1400	–
Metal coating	690	430	–
Centralized waste management	222–1320	–	160–283
Electronic components	720–1120	–	270–410

Table 10 Recommended mercury exposure limits

TDI RIVM (Baars et al. 2001) (ng/kg bw/day)	PTWI JECFA (2011) (ng/kg bw/week)	RfD (EPA 2021) (ng/kg bw/day)	MRL (ATSDR 2021) (ng/kg bw/day)
2000 (inorganic, oral) 100 (organic, oral)	4000 (inorganic) (WHO 2021a, b) 1600 (methylmercury) (WHO 2022)	100 (methylmercury, oral)	300 (methylmercury, chronic oral)

MRL minimum risk level

3.4 *Mercury*

Mercury (Hg) is a highly toxic, its uses are in products such as personal care products, thermometers, fluorescent light bulbs, electrical switches, pigments, batteries, and dental amalgams. Its usage is being phased out due to its adverse effects on human health (Crossett 2011; Suess et al. 2020).

3.4.1 Health Hazards

The health hazards caused by exposure of mercury depend upon its form (elemental, organic, or inorganic). Mercury can be accumulated in the human body and is also harmful to many biological systems, including the kidneys, the brain, and the epidermis. The elemental and organic forms of mercury can cross the blood–brain and placental barriers, accumulate in the brain, and develop a fetus (Ministry of Health 2021a, b). Mercury affects the kidneys, brain, skin and can accumulate in the brain, neurological effects, and developing fetus (ATSDR 1999; JECFA 2007). Table 10 represents different government agencies recommended exposure limits for mercury.

3.4.2 Sources of Industrial Effluent

Mercury in effluent is found from dental practice wastes (Bender 2008), papers, electrical utilities, and also from metal sectors (such as mining, primary, and fabricated metals).

3.4.3 Limit of Discharge and Policies

Many countries have recognized mercury as a hazardous substance which led Minamata Convention to diminish mercury emissions in the world (Suess et al. 2020). Under the ELGs, the US EPA has set mercury discharge limits for various point source categories (EPA 2021) summarized in Table 11. The daily maximum concentration of mercury in point source inorganic chemicals manufacturing is highest 110 μg/L to lowest in generating steam electric power (0.0018–0.788) μg/L, other contributors being centralized waste management, ore mining and dressing, etc.

3.5 *Arsenic*

Arsenic (As), a metalloid, was used in pesticides, pharmaceuticals, and agriculture-based industries (IARC Working Group on the Evaluation of Carcinogenic Risks to Humans 2012). Due to its toxic nature, many industries stopped its uses, though metal industry uses it as an alloying agent, leather industry uses during tanning of hides, and other uses is in manufacturing of paper, paint pigments, metal adhesives, glass, ammunition, and wood preservatives.

3.5.1 Health Hazards

Inorganic arsenic is highly toxic as compared to organic arsenic. Arsenic is known to be carcinogenic, lead to skin, bladder, and lung cancer. Moderate levels but long-term exposure damage heart, kidneys, liver, nerves and blood vessels (Ministry of

Table 11 US EPA mercury discharge limits summary

Point source	Daily maximum concentration (ng/L)	Average concentration [monthly] (μg/L)	Max average concentration [monthly] (μg/L)
Ore mining and dressing	2000	1000	–
Inorganic chemicals production	110,000	48,000	–
Waste combustors	2300	–	1300
Centralized waste management	641–17,200	–	246–6470

Table 12 Recommended arsenic exposure limits

TDI RIVM (Baars et al. 2001) (ng/kg bw/month)	RfD (EPA 2021) (ng/kg bw/day)	MRL (ATSDR 2021) (ng/kg bw/day)
30,000 (inorganic, oral)	300 (inorganic, oral)	300 (chronic, oral)

Health 2021a, b). Table 12 represents the different government agencies recommended exposure limits for arsenic.

3.5.2 Sources of Industrial Effluent

Arsenic is usually found as an impurity from metal ores and consequently enters the mining industry. Other potential sources of arsenic include the paper industry, the generation of steam-powered electricity, the refining of wood products, and waste treatment, among others.

3.5.3 Limit of Discharge and Policies

The US EPA has set limits for arsenic present in discharges as daily maximum concentration from different point source (US EPA ELGs 2021), as described in Table 13.

4 Per- and Polyfluoroalkyl Substances

Per- and poly-fluoroalkyl substances are a broad family of synthetic chemicals with resistance to water, oil, grease, and fire. These properties led to its widespread use in the production of numerous products, such as fabrics and carpets that are stain- and water-resistant, coatings, aviation hydraulic fluids, cleaning products, insect lures, firefighting foams, and in the electroplating and electronic industries all contain

Table 13 US EPA arsenic discharge limits summary

Sources	Maximum concentration [daily] (ng/L)	Average concentration [monthly] (ng/L)	Max average concentration [monthly] (ng/L)
Ore mining and dressing	1000	500	–
Inorganic chemicals production	3000	1000	–
Timber product processing	4000	–	–
Waste combustors	84	–	72
Centralized waste management	99.3–2950	–	19.9–1330
Landfills	1100	–	540
Electrical components	2090	830	–

polyurethane. Due to their resistance to degradation, these substances persist in the environment for extended periods of time and bioaccumulate in tissues. Among the 3000 PFAS, perfluoro octane sulfonate (PFOS) and perfluorooctanoic acid (PFOA) are the most well known (Wang et al. 2017). Several PFAS being phased out because of environmental concerns, Stockholm Convention also enlisted PFOS and PFOA.

These substances gets mixed with wastewater network from both industrial and residential sources which are most substantial (Rumsby 2018). The estimated half-lives of PFOS and PFOA in the human body are 3.4 and 2.7 years, respectively (Li et al. 2018), though PFOS and PFOA are not considered to pose acute health risks. Industrial sources for these compounds among others are the textile, metal plating industry, pulp and paper industry, semiconductor, and electronics business (Lin et al. 2009; Tonkin and Taylor Ltd 2018; Kim et al. 2021).

4.1 *Health Hazards*

The effects of PFOS and PFOA exposure are not considered to pose an acute health risk (Ministry for the Environment 2021). The chronic exposure limit is on serum cholesterol levels and immune effects, in spite of that the estimated half-lives of PFOS and PFOA in the human body are 3.4 and 2.7 years, respectively (Li et al. 2018), Table 14 represents the different government agencies recommended arsenic exposure limits.

4.2 *Sources of Industrial Effluent*

Textile industry is one of the major industries utilizing PFAS (Ministry of Environment and Food, The Danish Environmental Protection Agency 2015), in flame-retardant clothing and to impart water, oil, and dirt resistance into fabrics and carpets (Tonkin and Taylor Ltd 2018).

PFAS is commonly used in the metal plating industry (HRP Associates 2021), chromium electroplating facilities, pulp and paper industry, and semiconductor and electronics industries (Tonkin and Taylor Ltd 2018) resulted in contamination to wastewater (US EPA 2021a, b, c, d, e, f, g, h).

Table 14 Recommended PFOS and PFOA exposure limits

Substance	TWI (pg/kg bw/week) (EFSA 2022a, b)	TDI (pg/kg bw/day) (FSANZ 2022)	Draft RfD (pg/kg bw/day) (EPA 2021)
PFOS	4400*	20,000	7.9 (chronic oral)
PFOA		160,000	1.5 (chronic oral)

*A group TWI for PFAS based on PFOA, PFOS, perfluorononanoic acid (PFNA), and perfluorohexane sulfonic acid (PFHxS) assessments

Table 15 PFOS and PFOA discharge reported by the US Toxics Release Inventory 2020

Substances	Main industry linked to the facility	TRI industry sector	Surface water discharge (kg)[2]
PFOS	All other chemical preparation and product production	Chemicals	0.5
PFOA	All other chemical preparation and product production	Chemicals	4.1

4.3 Limit of Discharge and Policies

The US EPA developed effluent guidelines based on data acquired from a "preliminary multi-industry PFAS study" and mandated PFAS manufacturers to adhere to these standards (US EPA 2021a, b, c, d, e, f, g, h).The New Zealand EPA set PFOS and PFOA in waste discharge limits at 0.1 μg/L and 1 μg/L for total PFAS (Dawson 2018). The EPA also proposed limit for PFOS in biosolids to be at 0.3 mg per kg dry weight (Table 15).

5 Pharmaceuticals

It is not possible to individually assess the hazards posed by each industrial effluent due to the wide variety of pharmaceutical drugs available worldwide. Wastewater treatment plants in industries are noted to be poorly equipped (Orias and Perrodin 2013), and this becomes the major source of pharmaceuticals to the aquatic species (Larsson et al. 2007; Sengar and Vijayanandan 2022; ANSES 2013; Khetan and Collins 2007; WHO 2012). Also from human excretion, it entered municipal wastewater and groundwater.

5.1 Health Hazards

Several researches have determined the environmental and ecological effects of pharmaceuticals in aquatic environments (Khetan and Collins (2007), Orias and Perrodin (2013), and Orias and Perrodin (2014)), its impact on human is less known (Khetan and Collins 2007). It has been determined that very little amounts of pharmaceuticals in drinking water are extremely unlikely to pose health hazards to humans (WHO 2012), antibiotics contribution to the growth of antimicrobial resistance (AMR) may constitute a threat to human health (Kumar et al. 2019; Larsson et al. 2007; Sengar and Vijayanandan 2022).

5.2 *Sources of Industrial Effluent*

Pharmaceuticals entered municipal WWTPs through residential and trade waste route. Consumed or improperly discarded medications that are flushed down the commode or sink constitute residential contributions (WHO 2012), whereas industrial contributions consist of effluents containing pharmaceuticals from hospital waste, pharmaceutical company, and old care facilities (Orias and Perrodin 2014). Compared to residential wastewater, hospital effluents contain significantly higher concentrations of pharmaceuticals (Verlicchi et al. 2012; Majumder et al. 2021).

5.3 *Limit of Discharge and Policies*

There are guidelines relating to hospital wastewater management (WHO 2014; US EPA ELGs for Hospitals) without specific standards for pharmaceutical pollutants (Majumder et al. 2021). The US EPA has also set ELGs for the pharmaceutical manufacturers (US EPA ELGs 2022a, b, c) without mentioning discharge limits for specific pharmaceutical products. The Organization for Economic Cooperation and Development (OECD) 2019 published a report categorizing source-directed, use-oriented, and end-of-pipe policy instruments.

All these three categories are further subdivided into regulatory, economic, and voluntary. Regulations include environmental quality, good manufacturing practices, effluent discharge, best available techniques, and product bans. Economic include product or substance charges and subsidies. Lastly, voluntary includes advisory services, waste collection, public environmental health campaigns, disease prevention, and eco-labeling of green pharmaceuticals. It is directed to decrease or diminish the discharge of pharmaceuticals into water resources (OECD 2019) to protect drinking water sources.

6 Pesticides

As like pharmaceuticals, pesticides also exist in wide variety in numbers. It is not possible to assess individually the risk posed by each chemical in industrial effluents. They entered to wastewater through industrial effluent and from agricultural land and thus effect the aquatic environment.

Table 16 World Health Organization basis for pesticide classifications

	LD_{50} for the rat (gm/kg body weight)	
Class	Oral	Dermal
Ia extremely hazardous	<0.005	<0.05
Ib highly hazardous	0.005–0.05	0.05–0.2
II moderately hazardous	0.05–2	0.2–2
III slightly hazardous	Over 2	Over 2
U unlikely to present acute hazardous	5 or higher	

6.1 Health Hazards

Among the three primary categories of pesticides, only insecticides interfere with the nervous system, whereas fungicides and herbicides have a much wider range of potential health effects. The World Health Organization categorizes pesticides in accordance with their acute oral and dermal toxicity to rats (WHO 2020). These categories are extremely (Ia) and highly hazardous (Ib); moderately hazardous (II); slightly hazardous (III); and unlikely to present acute hazard (U) in normal use (WHO 2020), as summarized in Table 16.

6.2 Sources of Industrial Effluent

Significant quantities of pesticides have been detected in the wastewaters from multiple pesticide manufacturing facilities around the world (Affam et al. 2014, Pham et al. 2021). In addition, pesticides have been detected in effluent discharged from the agro-based industry, including fruits and vegetables (Campos-Mañas et al. 2019), and fruit-packaging industry (Karas et al. 2016).

6.3 Limit of Discharge and Policies

The US EPA has determined ELGs of a daily maximum of 0.01 kg of organic pesticide chemicals per 1000 kg of total organic active ingredients and a monthly average of 0.0018 kg of organic pesticide chemicals per 1000 kg of total organic active ingredients (US EPA ELGs 2022a, b, c). In Italy, the discharge limit of total pesticides is 50 μg/L (Mezzanotte et al. 2005), in New Zealand, the Model General Bylaw for Trade Waste (Standards New Zealand 2004) sets a maximum concentration of 0.2 g/m^3 (200 μg/L) for total pesticides discharge limit, and in Taiwan, discharge limits depend on types pesticides (Hamilton et al. 2003).

7 Microplastics

Microplastics refer to plastic smaller than 5 mm in length, it comprises various polymers and different chemical additives (Rochman et al. 2019). They emerge from the breakdown of larger plastic pollutant that comes from residential and industrial sources.

7.1 *Health Hazards*

Microplastics serve as vectors for toxic contaminants; they can be transported through circulatory system to distant sites in the body (Rahman et al. 2021). In addition, microplastics have also been detected in human placentas (Ragusa et al. 2021). Microplastics infiltrate the wastewater network in large quantities from several household sources, including significant amounts of microfibers released when washing synthetic clothing (Prata 2018).

7.2 *Sources of Industrial Effluent*

Large amounts of microplastics pass through wastewater network from several household sources, including washing of synthetic clothes in the form of microfibers (Prata 2018). Though treatment processes reduce microplastic concentration to low level, a high volume of effluents released daily add substantial amount to wastewater (Conley et al. 2019, Prata 2018, Sun et al. 2019, Conley et al. 2019). Wastewater-containing antibiotics get contact with microplastics and together they may develop extracellular antibiotic resistance genes and antibiotic-resistant bacteria (Syranidou and Kalogerakis 2021). Industrially, polymer processing plant (Bitter and Lackner 2020), textile manufacturing industry (Chan et al. 2021, Xu et al. 2018, Zhou et al. 2020), marine construction facilities (Franco et al. 2020), machine manufacturing, and chemical and electroplating plants in China (Wang et al. 2020) contribute microplastic to wastewaters.

7.3 *Limit of Discharge and Policies*

The European Commission is creating a microplastics initiative to decrease the accidental discharge of microplastics into the environment (European Commission 2021). Water Research Australia studied that effluents that contain microplastic are not currently regulated under discharge licenses (Water Research Australia 2021). In New Zealand, Aotearoa Impacts and Mitigation of Microplastics (AIM2) Ministry of

Business, Innovation and Employment Endeavour (MBIE) also investigates microplastics in wastewater (ESR 2021).

8 Conclusions

The objective of this study was to understand contaminants originating from industrial as well as municipal waste. This chapter tries to provide an international perspective on contaminants that are more concerned with human health determined in industrial effluents based on the literature. The selected contaminants were grouped into six classes, such as endocrine-disrupting compounds, heavy metals and metalloids, per- and polyfluoroalkyl substances, pharmaceuticals, pesticides, and microplastics. The industries generating these effluents were identified and potential health effects to human were also highlighted. Biodiversity and ecosystem services are also affected by these pollutants. Within the context of the mitigation hierarchy, environmental impacts should first be avoided, then minimized, and finally restored where possible. To reduce the adverse effect, treatment facilities, both onsite and off-site, need to installed. Factors causing the unrepaired environmental damage pay compensation, while the society does not bear the cost. The methods of environmental compensation and minimization of negative impacts on the environment are measures to protect environmental elements. This environmental compensation may relate to protecting plants, animals, or protection against pollution of air, water, and soil. The local body should also ensure that industries compensate the farmer for the loss of vegetation caused by the effluent.

References

Abduli MA (1996) Industrial waste management in Tehran. Environ Int 22:335–341

Addo K (2013) Solid waste management in Ghana: a case study of Effiduase and Asokore in the Sekyere East District, Master thesis, Kwame Nkrumah University of Science and Technology, Kumasi, Ghana

Affam AC, Chaudhuri M, Kutty SRM et al (2014) UV Fenton and sequencing batch reactor treatment of chlorpyrifos, cypermethrin and chlorothalonil pesticide wastewater. Int Biodeter Biodegr 93:195–201

Aivalioti M, Cossu R, Gidarakos E (2014) New opportunities in industrial waste management. Waste Manag 34(10):1737–1738

ANSES (2013) Health risk assessment associated with the presence of pharmaceuticals in drinking water: general method and application to carbamazepine and danofloxacin: expert report. French Agency for Food, Environmental and Occupational Health & Safety, Maisons-Alfort, France

Ashworth M, Chappell A (2015) Health risk assessment of selected phthalates reported from a local survey of children's plastic toys in Christchurch, New Zealand. Institute of Environmental Science and Research, Christchurch

ATSDR (1999) Toxicological profile for mercury. Agency for Toxic Substances and Disease Registry, Atlanta, GA

ATSDR (2021) Minimal risk levels (MRLs) for hazardous substances. Available at https://wwwn.cdc.gov/TSP/MRLS/mrlslisting.aspx
Baars AJ et al (2001) Re-evaluation of human-toxicological maximum permissible risk levels. Rijksinstituut voor Volksgezondheid en Milieu RIVM. RIVM Rapport 711701025
Balabanic D, Klemencic AK (2011) Presence of phthalates, bisphenol A, and nonylphenol in paper mill wastewaters in Slovenia and efficiency of aerobic and combined aerobic-anaerobic biological wastewater treatment plants for their removal. Fresen Environ Bull 20:86–92
Bender M et al (2008) Facing up to the hazards of mercury tooth fillings. In: A report to US House of Representatives Government Oversight Committee on Domestic Policy assessing state and local regulations to reduce dental mercury emissions. https://www.non-au-mercure-dentaire.org/_fichiers/submission_mercury_policy_project.pdf. Mercury 2008
Bergé A, Gasperi J, Rocher V, Gras L, Coursimault A, Moilleron R (2014) Phthalates and alkylphenols in industrial and domestic effluents: case of Paris conurbation (France). Sci Total Environ 488–489:26–35. https://doi.org/10.1016/j.scitotenv.2014.04.081
Bharagava RN, Saxena G, Mulla SI, Patel DK (2018) Characterization and Identification of Recalcitrant Organic Pollutants (ROPs) in Tannery Wastewater and Its Phytotoxicity Evaluation for Environmental Safety. Arch Environ Contam Toxicol 75(2):259–272. https://doi.org/10.1007/s00244-017-0490-x
Bitter H, Lackner S (2020) First quantification of semi-crystalline microplastics in industrial wastewaters. Chemosphere 258:127388
Buyukada M (2019) Removal potential reaction pathways and overall cost analysis of various pollution parameters and toxic odor compounds from the effluents of turkey processing plant using TiO2–assisted UV/O3 process. J Environ Manag 248:109298. https://doi.org/10.1016/j.jenvman.2019.109298
Campos-Mañas MC, Plaza-Bolaños P, Martínez-Piernas AB et al (2019) Determination of pesticide levels in wastewater from an agro-food industry: target, suspect and transformation product analysis. Chemosphere 232:152–163
Casares ML, Ulierte N, Matarán A, Ramos A, Zamorano M (2005) Solid industrial wastes and their management in Asegra (Granada, Spain). Waste Manag 25:1075–1082
CDC (2021) Phthalates factsheet. Available at https://www.cdc.gov/biomonitoring/Phthalates_FactSheet.html
Çetin D, Dönmez S, Dönmez G (2008) The treatment of textile wastewater including chromium(VI) and reactive dye by sulfate-reducing bacterial enrichment. J Environ Manag 88(1):76–82. https://doi.org/10.1016/j.jenvman.2007.01.019
Chan CKM, Park C, Chan KM et al (2021) Microplastic fibre releases from industrial wastewater effluent: a textile wet-processing mill in China. Environ Chem 18:93–100
CHAP (2014) Chronic hazard advisory panel on phthalates and phthalate alternatives. US Consumer Product Safety Commission, Directorate for Health Sciences, Bethesda, MD
Conley K, Clum A, Deepe J et al (2019) Wastewater treatment plants as a source of microplastics to an urban estuary: removal efficiencies and loading per capita over one year. Water Res X 3:100030
Cressey P (2018) Risks to public health from emerging organic contaminants in the New Zealand aquatic environment. Institute of Environmental Science and Research, Christchurch
Crossett A (2011) Mercury pollutant minimization plan: Gerorgia-Pacific LLC
Davidson J, Meyers D, Chakraborty M (1992) No time to waste: poverty and the global environment. Oxfam, Oxford
Dawson P (2018) Disposal of PFAS containing wastewater to trade waste: environmental protection authority
Demirbas A (2011) Waste management, waste resource facilities and waste conversion processes, energy convers. Manage 52:1280–1287
Dermentzis K, Christoforidis A, Valsamidou E et al (2011) Removal of hexavalent chromium from electroplating wastewater by electrocoagulation with iron electrodes. Global NEST J 13:412–418

Diamanti-Kandarakis E, Bourguignon JP, Giudice LC et al (2009) Endocrine-disrupting chemicals: an endocrine society scientific statement. Endocr Rev 30:293–342
Eaton C (2022) Review of potential health hazards associated with industrial effluents e Institute of Environmental Science and Research Limited (ESR). Ministry of Health
EFSA (2007) ONE Conference 2022. EFSA
EFSA (2014a) CONTAM. Available at https://efsa.onlinelibrary.wiley.com. https://doi.org/10.2903/j.efsa.2014.3595
EFSA (2014b) Scientific opinion on the risks to public health related to the presence of chromium in food and drinking water. EFSA Panel on Contaminants in the Food Chain (CONTAM), European Food Safety Authority (EFSA), Parma, Italy. EFSA J 12:3595
EFSA (2015) Annual Report of the European Food Safety Authority for 2015. Europa
EFSA (2018) Dioxins and related PCBs: tolerable intake level updated. Available at https://www.efsa.europa.eu/en/press/news/dioxins-and-related-pcbs-tolerable-intake-level-updated
EFSA (2021) EFSA sets lower tolerable intake level for cadmium in food. Available at https://www.efsa.europa.eu/en/news/efsa-sets-lower-tolerable-intake-level-cadmium-food
EFSA (2022a) Available at https://efsa.onlinelibrary.wiley.com. https://doi.org/10.2903/j.efsa.2019.5838
EFSA (2022b) PFAS in food: EFSA assesses risks and sets tolerable intake. Available at https://www.efsa.europa.eu/en/news/pfas-food-efsa-assesses-risks-and-sets-tolerable-intake
El Sidig NOA (2004) Solıd waste management in Khartoum industrial area, Master thesis, University of Khartoum, Sudan
Environment Canada and Health Canada (2001) Nonylphenol and its ethoxylates. Ottawa
EPA (2021) Arsenic, inorganic. Available at https://iris.epa.gov/ChemicalLanding/&substance_nmbr=278
ESR (2021) Aotearoa impacts and mitigation of microplastics (AIM^2). Available at https://www.esr.cri.nz/our-research/research-projects/aotearoa-impacts-and-mitigation-of-microplastics-aim/
European Commission (2002) European Union risk-assessment report vol 10, 2002 on 4-nonylphenol (branched) and nonylphenol. European Chemicals Bureau, Joint Research Centre, European Commission, Ispra, Italy
European Commission (2021) Microplastics. Available at https://ec.europa.eu/environment/topics/plastics/microplastics_en
Food Standards Australia New Zealand (2011) The 23rd Australian total diet survey. Food Standards Australia New Zealand, Canberra
Food Standards Australia New Zealand (2022) Perfluorinated compounds. Available at https://www.foodstandards.gov.au/consumer/chemicals/Pages/Perfluorinated-compounds.aspx
Franco AA, Arellano JM, Albendín G et al (2020) Mapping microplastics in Cadiz (Spain): occurrence of microplastics in municipal and industrial wastewaters. J Water Process Eng 38:101596
Fuerhacker M (2003) Bisphenol A emission factors from industrial sources and elimination rates in a sewage treatment plant. Water Sci Technol 47:117–122
Gautam GJ, Chaube R, Joy K (2015) Toxicity and tissue accumulation of 4-nonylphenol in the catfish heteropneustesfossilis with a note on prevalence of 4-NP in water samples. Endocrine Disruptors 3:e981442
Government of Canada (2017) Risk management scope for 1,2-benzenedicarboxylic acid, bis (2-ethylhexyl) ester [DEHP]. Available at https://www.ec.gc.ca/ese-ees/default.asp?lang=En&n=E00E9A1F-1
Government of Canada (2018) Bisphenol A. Available at https://www.canada.ca/en/environment-climate-change/services/evaluating-existing-substances/federal-environmental-quality-guidelines-bisphenol-a.html
Government of Canada (2021) Pulp and paper mill effluent chlorinated dioxins and furans regulations. Available at https://laws.justice.gc.ca/eng/regulations/SOR-92-267/FullText.html
Grisham JW (1986) Health aspects of the disposal of waste chemicals. Pergman Press, Oxford

Groshart CP et al (2001) Chemical study on alkylphenols. *Rijkswaterstaat*, RIKZ
Hamilton DJ, Ambrus Á, Dieterle RM et al (2003) Regulatory limits for pesticide residues in water - (IUPAC technical report). Pure Appl Chem 75:1123–1155
Hasheela R (2009) Municipal waste management in Namibia: the Windhoek case study, PhD thesis, Universidad Azteca, Mexico
Heindel JJ, Belcher S, Flaws JA, Prins GS, Ho SM, Mao J, Patisaul HB, Ricke W, Rosenfeld CS, Soto AM, vom Saal FS, Zoeller RT (2020) Data integration, analysis, and interpretation of eight academic CLARITY-BPA studies. Reprod Toxicol 98:29–60
Hinga KR, Batchellor A (2005) Waste processing and detoxification in Hassan RM, Scholes R, and Ash N eds., Ecosystems and human well-being: current state and trends Volume 1, Island Press, Washington, DC, p. 917
Ho H, Watanabe T (2017) Distribution and removal of nonylphenol ethoxylates and nonylphenol from textile wastewater—A comparison of a cotton and a synthetic fiber factory in Vietnam. Water 9(6):386. https://doi.org/10.3390/w9060386
HRP Associates (2021) PFAS in the metal plating industry fact sheet. Available at https://hrpassociates.com/uploads/files/Metal_Plating_Fact_Sheet.pdf?v=1623863694814#:~:text=PFAS%20have%20been%20used%20in,emissions%20of%20toxic%20metal%20fumes
IARC Working Group on the Evaluation of Carcinogenic Risks to Humans (2012) Arsenic and arsenic compounds in arsenic, metals, fibres and dusts. Volume 100 C. A review of human carcinogens. Lyon, France
JECFA (2007) Safety evaluation of certain food additives and contaminants. Prepared by the sixty-seventh meeting of the Joint FAO/WHO Expert Committee on Food Additives (JECFA). World Health Organization, Geneva
JECFA (2011) Safety evaluation of certain contaminants in food. In: Presented at Seventy-second meeting of the Joint FAO/WHO Expert Committee on Food Additives. JECFA, Rome, Italy
Kan A (2009) General characteristics of waste management: a review. Energy Educ Sci Technol Part A Energy Sci Res 23(1):55–69
Karas PA, Perruchon C, Karanasios E et al (2016) Integrated biodepuration of pesticide-contaminated wastewaters from the fruit-packaging industry using biobeds: bioaugmentation, risk assessment and optimized management. J Hazard Mater 320:635–644
Kawamoto K, Weber R (2021) Dioxin sources to the aquatic environment: re-assessing dioxins in industrial processes and possible emissions to the aquatic. Emerg Contam 7:52–62. https://doi.org/10.1016/j.emcon.2021.01.002
Khetan SK, Collins TJ (2007) Human pharmaceuticals in the aquatic environment: a challenge to green chemistry. Chem Rev 107:2319–2364
Kim KY, Ndabambi M, Choi S, Oh JE (2021) Legacy and novel perfluoroalkyl and polyfluoroalkyl substances in industrial wastewater and the receiving river water: temporal changes in relative abundances of regulated compounds and alternatives. Water Res 191:116830
Kumar R, Sarmah AK, Padhye LP (2019) Fate of pharmaceuticals and personal care products in a wastewater treatment plant with parallel secondary wastewater treatment train. J Environ Manage 233:649–659
Labunska I, Brigden K, Santillo D, Johnston P (2011) Heavy metal and organic chemical contaminants in wastewater discharged from leather tanneries in the Lanús district of Buenos Aires. greenpeace. https://www.greenpeace.to/greenpeace/wpcontent/uploads/2012/03/Argentina-tanneries-Technical-Note-07-2011-final.pdf
Larsson DGJ, de Pedro C, Paxeus N (2007) Effluent from drug manufactures contains extremely high levels of pharmaceuticals. J Hazard Mater 148:751–755
Lee HB, Peart TE (2000) Bisphenol A contamination in Canadian municipal and industrial wastewater and sludge samples. Water Qual Res J 35:283–298
Lee S, Liao C, Song GJ, Ra K, Kannan K, Moon HB (2015) Emission of bisphenol analogues including bisphenol A and bisphenol F from wastewater treatment plants in Korea. Chemosphere 119:1000–1006

Li J (2009) Types, amounts and effects of industrial solid wastes point sources of pollution: local effects and its control, vol I. EOLSS Publisher, Oxford
Li Y, Fletcher T, Mucs D, Scott K, Lindh CH, Tallving P, Jakobsson K (2018) Half-lives of PFOS, PFHxS and PFOA after end of exposure to contaminated drinking water. Occup Environ Med 75:46–51
Lin AYC, Panchangam SC, Lo CC (2009) The impact of semiconductor, electronics and optoelectronic industries on downstream perfluorinated chemical contamination in Taiwanese rivers. Environ Pollut 157:1365–1372
Majumder A, Gupta AK, Ghosal PS, Varma M (2021) A review on hospital wastewater treatment: a special emphasis on occurrence and removal of pharmaceutically active compounds resistant microorganisms and SARS-CoV-2. J Environ Chem Eng 9(2):104812. https://doi.org/10.1016/j.jece.2020.104812
Malakootian M et al (2009) Removal of heavy metals from paint industry's wastewater using Leca as an available adsorbent. Int J Environ Sci Technol 6:183–190
Mannetje A, Coakley J, Douwes J (2018) Report on the biological monitoring of selected chemicals of concern: results of the New Zealand biological monitoring programme, 2014–2016. Massey University, Wellington
Mezzanotte V, Canziani R, Sardi E et al (2005) Removal of pesticides by a combined ozonation/attached biomass process sequence. Ozone-Sci Eng 27:327–331
Ministry for the Environment (2021) PFOS and PFOA contamination: Engagement with potentially affected neighbours. Available at https://environment.govt.nz/site-search/?keyword=pfos%20%20and%20pfoa
Ministry of Environment and Food, The Danish Environmental Protection Agency (2015) Polyfluoroalkyl substances (PFASs) in textiles for children. Survey of chemical substances in consumer products. The Danish Environmental Protection Agency, Copenhagen
Ministry of Health (2020) Dioxins: a technical guide. Ministry of Health, Wellington
Ministry of Health (2021a) The environmental case management of mercury-exposed persons. Ministry of Health, Wellington
Ministry of Health (2021b) Arsenic and health. Available at https://www.health.govt.nz/your-health/healthy-living/environmental-health/hazardous-substances/arsenic-and-health#healtheffects
Mirzaee SA, Jaafarzadeh N, Gomes HT, Jorfi S, Ahmadi M (2019) Magnetic titanium/carbon nanotube nanocomposite catalyst for oxidative degradation of bisphenol A from high saline polycarbonate plant effluent using catalytic wet peroxide oxidation. Chem Eng J 370:372–386
Misra V, Pandey SD (2005) Hazardous waste, impact on health and environment for development of better waste management strategies in future in India. Environ Int 31:417–431
Muhammd BL et al (2018) Determination of cadmium, chromium and lead from industrial wastewater in Kombolcha town, Ethiopia using FAAS. Anal Chem
National Research Council (2008) Phthalates and cumulative risk assessment: the tasks ahead. Washington
NIEHS (2021) Bisphenol A (BPA). Available at https://www.niehs.nih.gov/health/topics/agents/sya-bpa/index.cfm
Nielsen E, Østergaard G, Thorup I, Ladefoged O, Jelnes JE (2000) Toxicological evaluation and limit values for nonylphenol, nonylphenol ethoxylates, tricresyl, phosphates and benzoic acid. Danish Veterinary and Food Administration
OECD (2019) Pharmaceutical residues in freshwater: hazards and policy responses. OECD Publishing, Paris. Available at https://www.oecd.org/environment/resources/Pharmaceuticals-residues-in-freshwater-policy-highlights-preliminary-version.pdf
Ojoawo S, Agbede O, Sangodoyin A (2011) On the physical composition of solid wastes in selected dumpsites of Ogbomosoland, South-Western Nigeria. J Water Resource Prot 3:661–666
Orias F, Perrodin Y (2013) Characterisation of the ecotoxicity of hospital effluents: a review. Sci Total Environ 454–455:250–276. https://doi.org/10.1016/j.scitotenv.2013.02.064

Orias F, Perrodin Y (2014) Pharmaceuticals in hospital wastewater: their ecotoxicity and contribution to the environmental hazard of the effluent. Chemosphere 115:31–39
Pappu A, Saxena M, Asolekar SR (2007) Solid wastes generation in India and their recycling potential in building materials. Build Environ 42:2311–2320
Pham TL, Boujelbane F, Bui HN et al (2021) Pesticide production wastewater treatment by electro-Fenton using Taguchi experimental design. Water Sci Technol 84:3155–3171
Pickston L, Brewerton HV, Drysdale JM et al (1985) The New Zealand diet: a survey of elements, pesticides, colours, and preservatives. NZ J Tech 1:81–89
Pothitou P, Voutsa D (2008) Endocrine disrupting compounds in municipal and industrial wastewater treatment plants in northern Greece. Chemosphere 73:1716–1723
Prata JC (2018) Microplastics in wastewater: state of the knowledge on sources, fate and solutions. Mar Pollut Bull 129:262–265
Ragusa A, Svelato A, Santacroce C et al (2021) Plasticenta: first evidence of microplastics in human placenta. Environ Int 146:106274
Rahman A, Sarkar A, Yadav OP, Achari G, Slobodnik J (2021) Potential human health risks due to environmental exposure to nano- and microplastics and knowledge gaps: a scoping review. Sci Total Environ 757:143872
Rochman CM, Brookson C, Bikker J et al (2019) Rethinking microplastics as a diverse contaminant suite. Environ Toxicol Chem 38:703–711
Rumsby A (2018) Poly and perfluorinated alkyl substances and wastewater treatment plants. Pattle Delamore Partners, Auckland
Sappington EN, Balasubramani A, Rifai HS (2015) Polychlorinated dibenzo-p-dioxins and polychlorinated dibenzofurans (PCDD/Fs) in municipal and industrial effluents. Chemosphere 133:82–89. https://doi.org/10.1016/j.chemosphere.2015.04.019
Sengar A, Vijayanandan A (2022) Human health and ecological risk assessment of 98 pharmaceuticals and personal care products (PPCPs) detected in Indian surface and wastewaters. Sci Total Environ 807:150677
Shang DY, Macdonald RW, Ikonomou MG (1999) Persistence of nonylphenol ethoxylate surfactants and their primary degradation products in sediments from near a municipal outfall in the strait of Georgia, British Columbia, Canada. Environ Sci Technol 33:1366–1372
Sise S, Uguz C (2017) Nonylphenol in human breast milk in relation to sociodemographic variables, diet, obstetrics histories and lifestyle habits in a Turkish population. Iran J Public Health 46:491–499
Soares A, Guieysse B, Jefferson B, Cartmell E, Lester JN (2008) Nonylphenol in the environment: a critical review on occurrence, fate, toxicity and treatment in wastewaters. Environ Int 34:1033–1049
Standards New Zealand (2004) Model general bylaw. Part 23—Trade waste (Standard No. 9201)
Stewart M, Northcott G, Gaw S, Tremblay LA (2016) An update on emerging organic contaminants of concern for New Zealand with guidance on monitoring approaches for councils: prepared by Streamlined Environmental Ltd, Northcott Research Consultants Ltd, University of Canterbury, Cawthron Institute and the University of Auckland for Auckland Council, Greater Wellington Regional Council and Environment Canterbury Regional Council
Suess E, Berg M, Bouchet S et al (2020) Mercury loads and fluxes from wastewater: a nationwide survey in Switzerland. Water Res 175:115708
Sun J, Dai X, Wang Q et al (2019) Microplastics in wastewater treatment plants: detection, occurrence and removal. Water Res 152:21–37
Syranidou E, Kalogerakis N (2021) Interactions of microplastics, antibiotics and antibiotic resistant genes within WWTPs. Sci Total Environ 804:150141
Tchobanoglous G, Theisen H, Eliassen R (1977) Solid wastes: engineering principles and management issues. McGraw Hill, Inc., New York
Tchobanoglous G, Theisen H, Vigil SA (1993) Integrated solid waste management: engineering principle and management issue. McGraw Hill Inc., New York

Tchounwou PB, Yedjou CG, Patlolla AK, Sutton DJ (2012) Heavy metal toxicity and the environment. Experientia Suppl 101:133–164
Tonkin and Taylor Ltd (2018) Scoping study: non fire-fighting foam sources of PFAS contamination in New Zealand: Report prepared for Environment Canterbury
U.S. Food and Drug (2021) Phthalates in cosmetics. Available at https://www.fda.gov/cosmetics/cosmetic-ingredients/phthalates-cosmetics#pht
US Environmental Protection Agency (2018) Final 2016 effluent guidelines program plan. US Environmental Protection Agency, Washington
US EPA (1984) Health assessment document for chromium. Environmental Assessment and Criteria Office, U.S. Environmental Protection Agency, Research Triangle Park
US EPA (1987a) Dibutyl phthalate (DBP). Available at https://iris.epa.gov/ChemicalLanding/&substance_nmbr=38
US EPA (1987b) Di (2-ethylhexyl)phthalate (DEHP). Available at https://iris.epa.gov/ChemicalLanding/&substance_nmbr=14
US EPA (1989) Butyl benzyl phthalate (BBP). Available at https://iris.epa.gov/ChemicalLanding/&substance_nmbr=293
US EPA (2021a) Risk management for bisphenol A (BPA). Available at https://www.epa.gov/assessing-and-managing-chemicals-under-tsca/risk-management-bisphenol-bpa
US EPA (2021b) 2,3,7,8-Tetrachlorodibenzo-p-dioxin. Available at https://iris.epa.gov/ChemicalLanding/&substance_nmbr=1024
US EPA (2021c) Chromium (III), insoluble salts. Available at https://iris.epa.gov/ChemicalLanding/&substance_nmbr=28
US EPA (2021d) Chromium (VI). Available at https://iris.epa.gov/ChemicalLanding/&substance_nmbr=144
US EPA (2021e) Bisphenol A. Available at https://www.epa.govt.nz/database-search/chemical-classification-and-information-database-ccid/view/B41EC9A7-D80E-41DF-A353-4AF0DCA80AC4
US EPA (2021f) Lead. Available at https://www.epa.gov/lead/learn-about-lead
US EPA (2021g) Methylmercury (MeHg). Available at https://iris.epa.gov/ChemicalLanding/&substance_nmbr=73
US EPA (2021h) Per- and polyfluoroalkyl substances (PFAS) Proposed PFAS National Primary Drinking Water Regulation. Available at https://www.epa.gov/sdwa/and-polyfluoroalkyl-substances-pfas/
US EPA ELGs (2021). Available at https://owapps.epa.gov/elg/
US EPA ELGs (2022a). Available at https://owapps.epa.gov/elg/results
US EPA ELGs (2022b) Guidance manual for pulp paper and paperboard and builders' paper and board mills pretreatment standards. Available at https://owapps.epa.gov/elg/
US EPA ELGs (2022c). Available at https://environment.govt.nz/what-government-is-doing/international-action/minamata-convention-on-mercury/
Vaiopoulou E, Gikas P (2020) Regulations for chromium emissions to the aquatic environment in Europe and elsewhere. Chemosphere 254:126876. https://doi.org/10.1016/j.chemosphere.2020.126876
Velusamy S, Roy A, Sundaram S, Mallick TK (2021) A review on heavy metal ions and containing dyes removal through graphene oxide-based adsorption strategies for textile wastewater treatment. Chem Rec 21:1570–1610
Verlicchi P, Al Aukidy M, Galletti A et al (2012) Hospital effluent: investigation of the concentrations and distribution of pharmaceuticals and environmental risk assessment. Sci Total Environ 430:109–118
Verma SK, Khandegar V, Saroha AK (2013) Removal of chromium from electroplating industry effluent using electrocoagulation. J Hazard Toxic Radioact Waste 17:146–152
Vigneswaran S, Jegatheesan V, Visvanathan C (1999) Industrial waste minimization initiatives in Thailand: concepts, examples and pilot scale trials. J Clean Prod 7:43–47

Wang JF (2018) Reuse of heavy metal from industrial effluent water. IOP Conf Ser Earth Environ Sci 199:042002

Wang Z, DeWitt JD, Higgins CP, Cousins IT (2017) A never-ending story of per- and polyfluoroalkyl substances (PFASs)? Environ Sci Technol 51:2508–2518

Wang F et al (2020) Occurrence and distribution of microplastics in domestic industrial agricultural and aquacultural wastewater sources: a case study in Changzhou China. Water Res 182:115956. https://doi.org/10.1016/j.watres.2020.115956

Water New Zealand (2017) Guidelines for beneficial use of organic materials on land. Water New Zealand, Wellington

Water Research Australia (2021) Occurrence, removal and risks of microplastics in drinking water and recycled water—State of knowledge. Available at https://www.waterra.com.au/research/open-rffs-and-rfps/2020/microplastics-in-wastewater-effluent/

Wei M-S, Huang K-H (2001) Recycling and reuse of industrial wastes in Taiwan. J Waste Manag 21:93–97

Wen X (2009) Point sources of pollution: local effects and their eontrol, vol I. EOLSS Publications, Oxford

Whittemore RC, LaFleur LE, Gillespie WJ, Amendola GA, Helms J (1990) USEPA/paper industry cooperative dioxin study: the 104 mill study. Chemosphere 20(10–12):1625–1632. https://doi.org/10.1016/0045-6535(90)90322-K

WHO (2003) General information related to microbiological risks in food. Available at http://www.who.int/foodsafety/micro/general/en

WHO (2012) Pharmaceuticals in drinking water. World Health Organization, Geneva

WHO (2014) World health statistics 2014. WHO

WHO (2019) Exposure to dioxins and dioxin-like substances: a major public health concern. Available at https://cdn.who.int/media/docs/default-source/food-safety/dioxins.pdf?sfvrsn=4bcd5f4d_1

WHO (2020) The WHO recommended classification of pesticides by hazard and guidelines to classification, 2019 edn., World Health Organization, Geneva

WHO (2021a) Cadmium. Available at https://apps.who.int/food-additives-contaminants-jecfa-database/chemical.aspx?chemID=1376

WHO (2021b) Mercury. Available at https://apps.who.int/food-additives-contaminants-jecfa-database/chemical.aspx?chemID=1806

WHO (2022) Methylmercury. Available at https://apps.who.int/food-additives-contaminants-jecfa-database/chemical.aspx?chemID=3083

WHO IPCS (2013). Available at https://apps.who.int/iris/handle/10665/90560

Wilbur S, Abadin H, Fay M et al (2012) Toxicological profile for chromium. Agency for Toxic Substances and Disease Registry, Altlanta

Xu X, Hou QT, Xue YG et al (2018) Pollution characteristics and fate of microfibers in the wastewater from textile dyeing wastewater treatment plant. Water Sci Technol 78:2046–2054

Yoshizawa S, Tanaka M, Shekdar AV (2004) Global trends in waste generation. In: Gaballah I, Mishar B, Solozabal R, Tanaka M (eds) Recycling, waste treatment and clean technology, TMS mineral. Metals and Materials Publishers, Spain, pp 1541–1552

Zhou HJ, Zhou L, Ma KK (2020) Microfiber from textile dyeing and printing wastewater of a typical industrial park in China: occurrence, removal and release. Sci Total Environ 739:140329

Zoeller RT, Bergman A, Becher G, Bjerregaard P, Bornman R, Brandt I, Iguchi T, Jobling S, Kidd KA, Kortenkamp A, Skakkebaek NE, Toppari J, Vandenberg LN (2014) A path forward in the debate over health impacts of endocrine disrupting chemicals. Environ Health 13:118

Zubair Alam M, Ahmad S, Malik A, Ahmad M (2010) Mutagenicity and genotoxicity of tannery effluents used for irrigation at Kanpur India. Ecotoxicol Environ Saf 73(7):1620–1628. https://doi.org/10.1016/j.ecoenv.2010.07.009

Zurbrugg C (2002) Solid waste management in developing countries. SANDEC/EAWAG

Accumulation of Heavy Metals in Roadside Plants and Their Role in Phytoremediation

Dipak Kumar Mahida, Vishal M. Makwana, Mahipal Singh Sankhla, Ankita Patel, and Pravinsang Dodia

Abstract Plants play very crucial roles in pollution control. According to their physical and chemical properties, contaminants can either be stable or labile. The movement of the stoma, or mouth of the leaf, and trichome adsorption are what allow plants to carry out the absorption process (spines or leaf hair). Heavy metals are the group of inorganic chemical pollutants and road traffic emissions that are most harmful to the biosphere. Since they cannot be broken down through biological and chemical processes, unlike organic pollutants, they tend to accumulate in the environment. Vehicular emissions and industrial exhausts harm the ecosystem while also causing heavy metal contamination. Untamed plants growing beside roadsides may be able to assist reduce heavy metal pollution. Heavy metals that impact the morphological, physiological, and reproductive characteristics of plants progressively change the pH of the soil. Roadside vegetation's germination and seedling development are impacted by heavy metal pollution. Phytoremediation can be employed as an alternative solution for heavy metal remediation processes because of its advantages as a low-cost, high-efficient, environmentally acceptable and eco-friendly techniques based on the utilization of metal accumulating plants. Future research on the number of heavy metals in a range of tropical roadside plants is necessary to determine the exact source and transport processes.

All the authors have contributed equally.

D. K. Mahida · A. Patel (✉)
Department of Biochemistry and Forensic Science, Gujarat University, Ahmedabad, Gujarat, India
e-mail: ankitapatel@gujaratuniversity.ac.in

V. M. Makwana · P. Dodia
Zoology Department, Sir P. P. Institute of Science, M. K. Bhavnagar University, Bhavnagar, Gujarat, India
e-mail: ppdodia@sirppscience.edu.in

M. S. Sankhla
Department of Forensic Science, Vivekananda Global University, Jaipur, Rajasthan, India

P. Pathak et al. (eds.), *Anthropogenic Environmental Hazards*,
https://doi.org/10.1007/978-3-031-41013-0_6

Keywords Roadside plants · Heavy metals · Pollution · Plant health

1 Introduction

Heavy metals may assemble and spread in the soil ecosystem. Plants vascular system and roots may allow them to take in metal pollution from the soil. Metal build-up in soil has the potential to threaten humans, plants, and animals as well as the ecosystem's stability. The capacity of the plant to synthesize chlorophyll may be hindered, oxidative stress may be increased, and stomata resistance may be weakened by high metal concentrations in the plant (Ashraf et al. 2011; Sulaiman 2018). Whether the pollution is in the soil or the air, unnaturally occurring heavy metals like chromium (Cr) and cadmium (Cd) can inhibit plant development. It is possible that heavy metals can enter people's bodies through the food chain, causing an increase in chronic illnesses like cancer and harming the central nervous system, especially in young children (Sulaiman 2018).

It is well acknowledged that vehicle traffic is a significant and growing source of air and soil pollution near highways. Since people are more likely to live close to vehicle exhaust than other types of air pollution (for example, along busy roads), they are more likely to breathe it in and come into touch with it. This makes vehicle pollution more dangerous than other types of air pollution. As a result, it is now widely acknowledged that transportation poses a serious threat to the health of the environment. Despite the fact that transportation emissions per vehicle have successfully decreased thanks to technology, more research is still required to develop a comprehensive understanding of all emissions and their effects on both the environment and human health. Numerous contaminants have been discovered to have greater concentrations in soil, water, and plants close to highways than farther away (Khalid et al. 2018).

Due to the lack of regulation governing nonexhaust pollution sources, attention to the effects of roadside dust to air pollution and human health has increased (Gope et al. 2018). Soil components, car exhaust and nonexhaust emissions, atmospheric deposition, and industrial activities are the main sources of heavy metals in roadway dust particles; among these, vehicle emissions are the major source of metals in metropolitan areas. Complex metal combinations from tire wear, brake and component wear, and resuspended road dust are only a few of the car emissions (Al-Taani et al. 2019). Plants may deposit and bioaccumulate heavy metals that are in the air and dust, making biomonitoring that uses plants a useful technique for evaluating environmental impact in urban settings. Because of their poisonous effects, propensity to bioaccumulate throughout the food chain, and high persistence in the environment, heavy metals are considered to be the principal category of inorganic pollutants. Heavy metal environmental pollution in various compartments has significantly grown globally, particularly in emerging nations (Fang et al. 2021).

In this chapter, authors have discussed various aspects of heavy metals accumulation in the roadside plant along with their hazard effects and mitigation strategies. Roadside use of native plants is a unique resource whose use goes far beyond the roadside, providing a toolbox for a fresh aesthetic that can be used on all sorts of public and private land. It is a crucial process for anybody interested in growing or restoring native vegetation and can help pave the way for an economical ecological approach to maintaining human-designed landscapes.

2 Heavy Metals in the Environment

Heavy metals are poisonous and highly reactive at low concentrations, polluting large regions of the world and posing serious dangers to human and environmental health. Even though many heavy metals are naturally present in the Earth's crust and atmosphere, human activities like mining, smelting, transportation, military operations, industrial manufacturing, and the use of metal-containing pesticides and fertilizers in commercial agriculture can promote heavy metal pollution. Through trash disposal, runoff, and the application of chemical products containing heavy metals, these activities release heavy metals into the environment, which may then enter terrestrial ecosystems through airborne deposition, surface waters, or soil (Gall et al. 2015). Heavy metals cannot be decomposed, in contrast to their organic pollutant counterparts. Heavy metals therefore remain in the environment for years after point sources of contamination have been eliminated (Babin-Fenske and Anand 2011).

2.1 Spatiotemporal Variations of HMs

Most scientists have studied how alterations in vegetation composition and structure are related to increased levels of nitrogen emissions. Elevated levels of atmospheric N can affect community diversity by favoring nitrophilous species (Bobbink et al. 1998). Early research during 1971/72 and 1990 showed that the prevalence of species that require nitrogen as well as the increase in the Ellenberg's ecological value N (nitrogen) were characteristics of the herb-layer vegetation in the Lorraine Plain (northeast of France), which was probably brought on by atmospheric deposition. A similar trend was noticed in Germany (Bernhardt-Römermann et al. 2007) or in Poland (Dzwonko 2001). Automobiles are one of the main contributors to heavy metal contamination in roadside soils. These pollutants enter the environment as particles from exhaust fumes or other vehicular components, eventually contaminating soil and vegetation in roadways and surrounding regions (Bohemen and van Janssen Van De Laak 2003). The most frequent heavy metals found in automobile exhaust include Cd, Cu, Pb, Ni, and Zn (Li et al. 2001; Elik 2003). The primary sources of Pb include domestic and industrial effluents, industrial paints, and exhaust emissions from gasoline-powered vehicles (Harrison and Laxen 1977). Despite the elimination of Pb from gasoline and the dramatically reduced amount of Pb entering sediments, the threats posed by previously deposited Pb in the environment are still very high. Cd is one of the most hazardous metals in the environment, easily transferring from soil to plants (Jia et al. 2010) due to its high solubility and mobility (Pagotto et al. 2001). Cd is typically found in automobile tyres and is contributed to sediments by the vehicle wear and the traffic propagation. For Cd and Pb uptake in all age categories, the World Health Organization considers 0.007 and 0.025 mg/kg, respectively, to be tolerable weekly limits (Bakirdere and Yaman 2008). On the Razan-Hamedan highway in Iran, Safari Sinegani (2007) studied the spatiotemporal fluctuations of Pb on the canopy of the *Amaranthus retroflexus* plant. The findings showed that Pb levels in plant samples were higher than allowed due to heavy traffic, the age of the road, seasonal variations, and distance from roadways. Roadside plants are especially susceptible to air pollution (Sinegani 2007; Lee et al. 2012). According to Princewill-Ogbonna and Ogbonna (2011), the results of a study conducted in Nigeria, soil and medicinal plants in Aba city have high accumulations of heavy metals, mostly due to vehicular emissions (Princewill-Ogbonna and Ogbonna 2011). Feng et al. (2012) found that traffic on the roadsides of rice and wheat fields in eastern China is a contributing factor to the significant rise in heavy metal levels in soil, rice, and wheat grains (Feng et al. 2012). Naser et al. (2012) concluded that as the distance from the road increased, the concentration of Pb and Ni in the soil and vegetables decreased in Bangladesh (Naser et al. 2012). However, concentration of Cd is independent from the distance to roads. Al-Chalabi and Hawker (2000) investigated the concentration of Pb around the roads in Brisbane, Australia and concluded that the primary source of pollution was traffic and Pb accumulation was noticed in the top 5 cm of the soil (Al-Chalabi and Hawker 2000). Fakayode and Olu-Owolabi (2003) investigated heavy metal concentration on the sidewalks of the streets of Osgbo, Nigeria and concluded that the concentration of heavy metals in the soil significantly reduced as the distance from the roadways increased up to 50 m (Fakayode and Olu-Owolabi 2003). Vehicle emissions pose serious risks to people, properties, and the surrounding biodiversity, including plants in metropolitan areas (Sánchez-Chardi 2016; Singh et al. 2020). In southwestern Nile Delta in Egypt, Khalifa and Gad (2018) collected samples of agricultural soils to investigate the heavy metal contamination (Khalifa and Gad 2018). The result indicated that the most significant causes of heavy metal contamination in the area were high urbanization, industrial activity, and agricultural activity.

2.2 Sources of HMs in the Environment

In general, heavy metals decrease enzyme activity and soil microbial biomass (SMB) levels which in turn reduces diversity of soil ecosystem and modifies microbial structure. There are several pathways through which heavy metals may enter the soil including the application of treated and untreated sewer sludge, runoff from

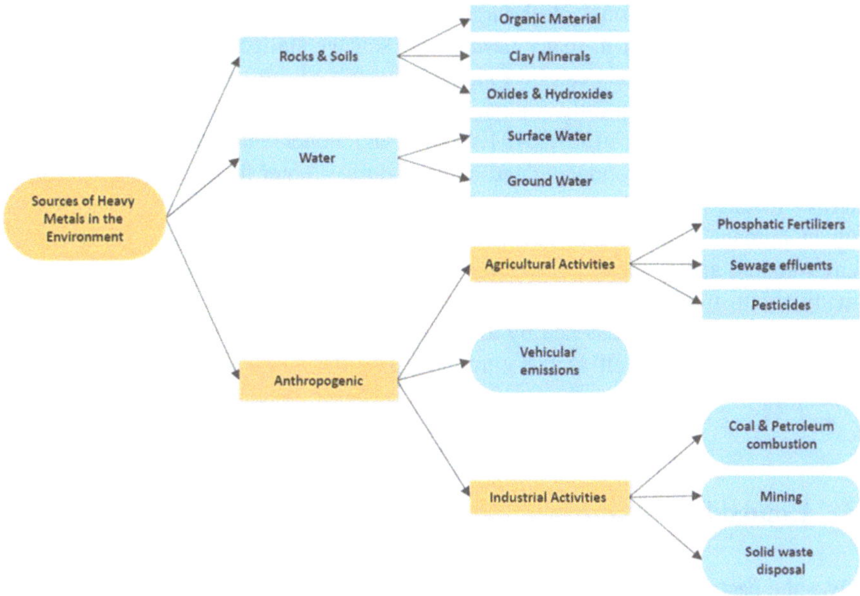

Fig. 1 Sources of heavy metals in the environment

industrial sites and highways, metallo-pesticides, phosphate fertilizers, and atmospheric deposition of metal-containing particles, refer Fig. 1. Heavy metals in plants can be extremely toxic to animals, leading to a variety of sublethal consequences including mortality. Invertebrates may inadvertently consume heavy metals found in soil and plants or absorb them through their outer body coverings or exoskeleton (Gall et al. 2015). When invertebrates consume metals, they may transfer those metals to their predators, particularly if those metals are present in the alimentary tract at the time of predation or accumulated in the tissues. Similar to small mammals, large mammals frequently consume contaminated vegetation (van der Fels-Klerx et al. 2011), small amounts of soil, or drinking water and they may also be exposed to metals by inhalation or grooming (Miranda et al. 2009). Heavy metal exposure in humans can occur in several ways, such as through dietary intake, workplace exposure, or environmental exposure through inhaling dust that contains contaminated particles (Amaya et al. 2013). The mining and milling of phosphate rock are significant sources of heavy metal air and soil pollution, in addition to the fact that phosphate-based fertilizers are a source of heavy metals in agricultural systems.

The total Cd input in Europe is 1.6 g ha^{-1} $year^{-1}$, with phosphate fertilizers accounting for 55% of the Cd and air deposition for 40% (Pan et al. 2010). According to Chauhan (2010), the amount of air pollution in India's urban areas caused by atmospheric dust pollutants is approximately 40% (Avnish 2008; Chauhan 2010). Sewage sludge is a by-product of sewage treatment that is produced in enormous quantities in China (30 million tonnes), Japan (70 million tonnes), and the USA (six million tons). With large amounts of available sewage sludge and the necessity to dispose of it, many farmers use sewage sludge to fertilize and irrigate farmlands. Heavy metals can be found in treated sewage sludge, which can come from sources like domestic inputs and wastewater, road runoff, and industrial contributions (Gall et al. 2015). Essential and nonessential heavy metals, both originating from anthropogenic and natural sources, enter food webs through soils. Heavy metals from soils may enter plants, insects, and grazers, and may then build up in humans, creating serious health hazards. To monitor heavy metals in the environment and evaluate the possibility of metal transmission within a food web, many organisms can be used (Stankovic et al. 2014). The health of both humans and animals is seriously threatened by the heavy metal contamination of plants and soil along roadways, as well as the introduction of these contaminants into the food chain.

3 Roadside Proximity

Over the past 40 years, several studies have been conducted on the pollution of the roadside environments along regularly used main roads and highways. The effect of restrictions like the prohibition on leaded gasoline, which has decreased emissions from individual automobiles, has been countered by rising global traffic. Road salts, metals, and polycyclic aromatic hydrocarbons, which can be found in both particulate and dissolved forms, are just a few of the toxins that are released from major roadways. Because they cannot be broken down by microorganisms (persistence) and have long-term toxicity for plants, animals, and people, metals in particular are a major problem. Cd, Cr, Cu, Pb, Ni, and Zn are the metals that are most frequently recognized and investigated in roadside situations. However, there are also higher concentrations of other metals, including As, Co, Sb, Se, Sr, and V, along regularly used roadways. Sb has received a lot of attention as a roadside pollutant in several recent studies due to its presence in vehicle brake pads, in particular (Pekey et al. 2004). In sense of roadside proximity, researchers are focusing on two main things: (1) roadside plants and (2) roadside soil. Additionally, there is a significant correlation between metal concentrations in flora and soil along roadsides.

The original physical, biological, and chemical characteristics of the soil in the nearby region are altered by the building, usage, and maintenance of roadways. The topsoil was frequently removed during road building or buried more than 1 m beneath the foundation course. Roadside soils frequently include up to 30% of stones and technologically derived minerals. These substances, together with other elements like alkaline deposition from road surfaces, cause soil pH to rise even higher than 7. The embankment, which was constructed when the road was being built, frequently measures up to 5 m and is situated very next to the edge of the road. There are frequently hills and ditches in this region as well to drain and infiltrate the runoff from the road. The soils are frequently disturbed and compacted at a distance of 5–10 m, with little to no vegetation. After this distance, the road's effect gradually fades, and original soil profiles predominately appear after 10–15 m. Many studies have been done to estimate the metal concentrations along European main roads and highways since roadside soils are one of the key targets for pollutants discharged from vehicles (Pekey et al. 2004).

The ability of many plant species to absorb, detoxify, and endure greater amounts of heavy metal contamination is well recognized. Few tree species are utilized to reduce air pollution, including *Mangifera indica L., Pongamia pinnata L. Pierre, Dalbergia sissoo Roxb,* and *Holoptelea integrifolia L.* Roadside plants often have higher levels of heavy metal tolerance, indicating their suitability for planting in ecological regions for the restoration of urban ecosystems. Urban wild plants are important for the ecosystem because they absorb air pollutants and lessen the island's heat-related effects. Therefore, trees and plants may help reduce air pollution and slow the effects of global warming (Altaf et al. 2021).

4 Phytoaccumulations of HMs by Roadside Plants

Phytoaccumulation is the process through which plants take up pollutants. In phytoaccumulation, plants or algae are primarily used to convert pollutants from soils, sediments, or water into harvestable plant biomass. More commonly than for organics, this method has been explored for heavy metal extraction. Until it is harvested, a live plant may continue to absorb toxins. In order to achieve considerable cleaning, the growing cycle must often be repeated through numerous crops because a lesser amount of contamination will still be present in the soil after harvesting.

Roadside vegetation is particularly vulnerable to air pollution, which includes a variety of potentially hazardous chemicals carried by particles. Some of these substances, including heavy metals and polyaromatic hydrocarbons, have an impact on plants by producing reactive oxygen species (ROS), refer Fig. 2 (Kováts et al. 2021).

4.1 Sensitivity of Roadside Plants

The diverse variety of symptoms displayed by plants in contaminated environments can even be used to evaluate the habitat quality in its whole. Physiological and morphological features, leaf water status, leaf ultrastructure, nonenzymatic and enzymatic antioxidants, photosynthetic pigments, nutritional status, and resource utilization are among the functional traits that have been discovered to be the most vulnerable to environmental stress (Mukherjee and Agrawal 2018). Many researchers have reported the bioindication of sensitivity toward air pollution in various plants which are described in Table 1.

However, it is important to emphasize that neither the precise exposure nor the quantity of the pollutant is known in the majority of field bioindication

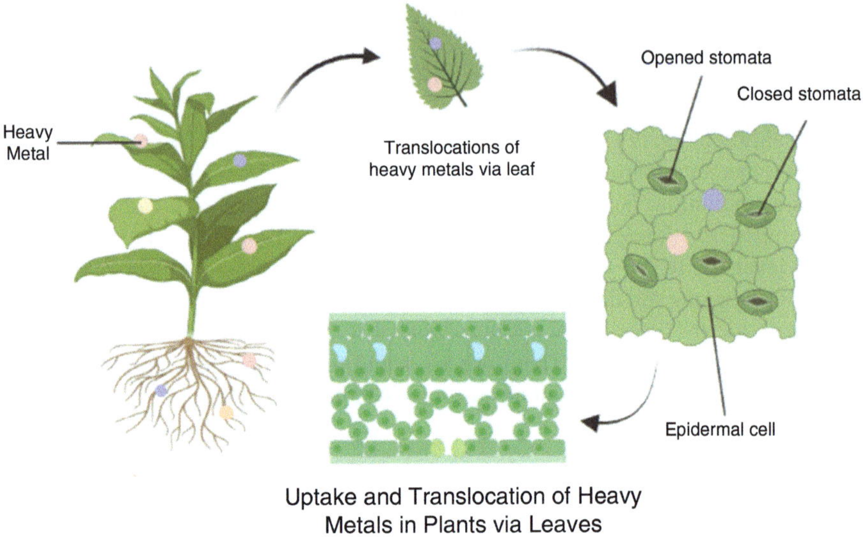

Fig. 2 Phytoaccumulation of HMs by leaf

Table 1 Sensitivity and bioindication in plants toward HMs

S. N.	Physiological/ morphological characteristics	Bioindications
1.	Stomatal conductance or stomatal density	Stomatal density increases and size decreases in stressed plants
2.	Leaf morphology	Decrease in specific leaf area (SLA) and enhanced leaf senescence
3.	Leaf injury	Necrotic spots, chlorotic patches like symptoms are seen
4.	Surface wax layer	The presence of lipophilic aromatic hydrocarbons in the emission was linked to the surface wax breakdown caused by exhaust gas exposure
5.	Biochemical markers	Ascorbic acid is one of the main compounds used in the defence against reactive oxygen species (ROS), hence greater ascorbic acid levels represent higher pollution loads. Also, some enzymes (catalase or peroxidase), antioxidant, and nonenzymatic antioxidant have similar pattern toward the pollution
6.	Photosynthesis activity	Stress can be quantified by the chlorophyll A to chlorophyll B ratio. A higher score denotes an improved tolerance for air pollution
7.	Photosynthetic pigments	A reduction in the protein content of the leaves is another sign of air pollution from contaminated environments

investigations. Furthermore, synergistic effects may cause symptoms to appear. Even though atmospheric deposition is now often blamed for many changes in vegetation, dose–effect, or more accurately, concentration-effect correlations are typically poorly understood (Kováts et al. 2021). To assess the impact of air quality on plants, a biological metric known as the Air Pollution Tolerance Index (APTI), is of major importance. The capacity of plants to mitigate the impacts of air pollution is an indication of their intrinsic qualities. High-index plants are typically tolerant of air pollution. They may live off of contaminated air. Low-index plants exhibit susceptibility to air pollution. Air contaminants are continuously exposed to by plants, which causes a build-up in their system. It changes the characteristics of the leaves and increases their sensitivity to contaminants. Because they make plants more sensitive, pollutants from traffic emissions endanger plant life (Sadia et al. 2019).

4.2 *Translocation of HMs in Plants*

In roadside plants, heavy metals mainly uptake through foliar way because of higher air pollution. However, only the root system has been studied in studies on the absorption and accumulation of heavy metals by crops and vegetables. This is because the bulk of heavy metals accumulate in soil systems and are mostly absorbed by plants through their roots. Heavy metals can also be absorbed by plant aerial organs including leaves, fruits, and flowers in addition to plant roots. According to this, plant aerial organs are effective absorbing organs that have processes comparable to the roots for absorbing heavy metals. After air particles have been deposited on the leaf surfaces, foliar transfer allows metals to accumulate in plant leaves. Increased foliar concentrations of heavy metals are seen in plants growing close to mining and smelting sites as well as in populated regions. Therefore, investigations on biomonitoring near businesses or roadways are presently given more consideration in order to assess metal pollution caused by air deposition or transmission (Shahid et al. 2016).

Foliar metal uptake can be described in two major steps:

1. Adsorption and internalization though the cuticle.
2. Penetration of metals though the stomatal pores.

About 300 years ago, it was discovered that plant foliar portions had the ability to absorb nutrients, water, and metals. The foliar transmission of metals can be overlooked or, conversely, seems to be the primary conduit of pollution, particularly when ultrafine particles interact with plant leaves. An effective filter for the emissions of heavy metals into the atmosphere is the plant canopy. Through stomata, cuticular cracks, lenticels, ectodesmata, and aqueous pores, heavy metals are absorbed by foliar surfaces, refer Fig. 3. In actuality, ectodesmata—no plasmatic channels found mostly between subsidiary cells and guard cells in the cuticular membrane or epidermal cell wall—are the primary pathways via which foliar deposited heavy metals are absorbed. Additionally, compared to epidermal cells, the cuticle located above the guard cell is relatively more porous (Shahid et al. 2016).

The cuticle is regarded as the initial barrier to the entrance of heavy metals because it is water repellent and acts as a protective layer. This layer was essential

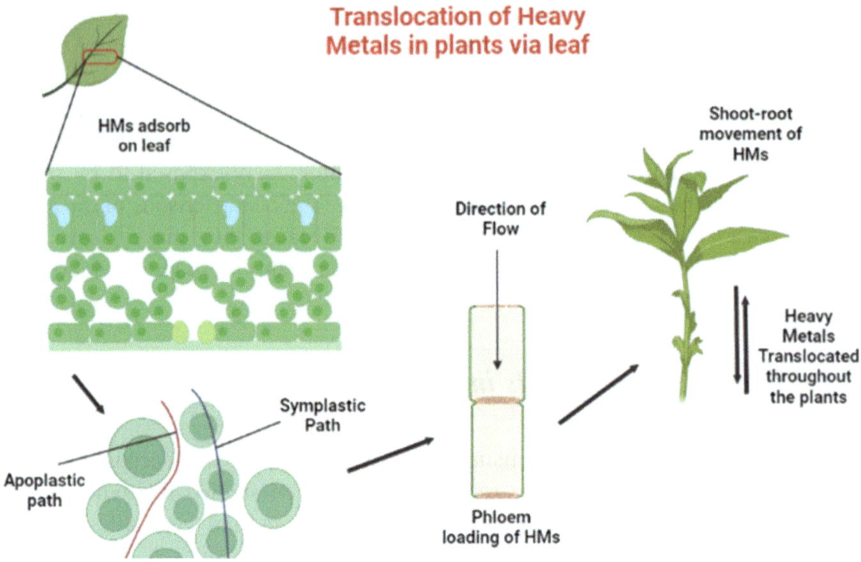

Fig. 3 Translocation of HMs in plant

for the defense system that kept water from evaporating, shielded plants from invaders, and aided in the flow of solutes. Additionally, there is little evidence of heavy metals penetrating through the cuticle. It is also important to keep in mind that cuticles will fluctuate according to various climatic conditions, life phases, and pest illnesses. Therefore, under some circumstances, the initial barrier (cuticle) to the translocation of heavy metals may be avoided.

Numerous scientific studies have described the stomatal pathway for the translocation of heavy metals in addition to the cuticular pathway. On leaves, stomata are microscopic holes that aid in gas exchange. It is interesting to note that the stomatal holes are larger than 20 nm in diameter. Then, through stomatal pathways, heavy metals are absorbed and move from leaves to interior tissue. Previous investigations have shown that heavy metals accumulate following foliar application in different plants' stomatal pathways, where they pass through. In addition, various plants have varying leaf morphologies and stomatal sizes, which affect the ability of heavy metals absorption. The thorough investigation of the stomatal route employed by various plant species to absorb heavy metals should be the main goal of future research. Such particles travel through the circulatory system after ingesting heavy metals through the stomatal route. The phloem system in plants allows different elements to go from shoot to root by following macromolecules, proteins, or RNA. Materials are often transported across long distances in plants by the vascular system, which is made up of the xylem and phloem. While the liquid in the phloem flows downhill from the shoot to roots, the elements of xylem go upward from the root to the plant's aerial portion.

The fact that the vascular system is thought of as no circulatory further establishes that once particles have moved from shoot to root, they do not return to their original location. The optimal method for heavy metals to migrate from upward to downward after foliar application is the phloem pathway, in order to sum up. However, it is still unclear how heavy metals are transported between the xylem and phloem. Another element that has a substantial impact on how plants absorb heavy metals through their foliar tissues is the phyllo sphere. Microorganisms that are symbiotic and harmful live in the phyllo sphere, which emits chemical signals. These exudates create a shield on the surface of the leaves that prevents the absorption of heavy metals. Furthermore, diseases like chlorosis, necrosis, leaf mould, and blight have a significant impact on this protective layer. As a result, the exudate layer on the leaf may become ineffective as a barrier, allowing heavy metals to enter.

4.3 *Adaptation and Mitigation Potential*

The first line of defence against urban vehicular pollution is the roadside planting. As well as absorbing gaseous pollutants, the surface area of the leaves allows for the settling of dispersed particulate matter. But eventually, individuals experience varying degrees of stress as evidenced by the physiological reaction. Most trees have an instantaneous physiological response to airborne contaminants before the appearance of the visual indication on leaves. The ability of the trees to withstand these pressures determines their capacity for adaptation and mitigation. A number of elements, such as the architecture, morphology, physiology, and biochemistry of plant tissues, influence the trees' ability to mitigate climate change. The effects of vehicle motion on photosynthesis, transpiration, stomatal conductance, proline content, and heavy metal concentration in plant tissue. The photosynthetic pigment is negatively impacted by air pollution, which lowers production. Air pollution has been reported to negatively affect leaf proteins, proline, carotene, and chlorophyll. Stomatal function and leaf thickness of urban plantations are affected by air pollution. One of the most reliable markers of stress is ascorbic acid, which has been linked to air pollution levels.

For many tree species, the ability to reduce air pollution is yet unknown, and it is unclear how each species would react in different situations. There have been reports of several biochemical, morphological, and anatomical alterations in trees growing beside roadsides as a result of greater air pollution concentrations. Some trees, nevertheless, have been shown to adjust their physiological and biochemical characteristics to survive larger doses. The more delicate tree varieties serve as biological markers of air pollution. Some trees can withstand quite high pollution levels and be utilized to efficiently reduce air pollution. Thus, it is critical to find trees that might reduce air pollution in urban settings and screen out trees that can serve as signs of stress caused by air pollution (Singh et al. 2020).

Air pollution and noise emissions have increased as a result of the growing population and number of cars nearby. It has been discovered that the neighboring people and tree species in that region are quite concerned about the pollution generated by autos. The roadside forest fragments, which are made up of different tree species, might in this way significantly lessen the impacts of air pollution brought on by automobiles and thereby lessen the enormous problem of air pollution. These remnants of the roadside woodland absorb carbon during photosynthesis, acting as a sink for atmospheric CO_2. Additionally, the forest remains beside the roadway and serve as a trap for gaseous and particle contaminants, including heavy metals. In general, heavy metal-containing particulate matter and gaseous pollutants are released into the atmosphere by industrial processes and motor movement. For those who are more prone to stay near vehicle exhaust, contaminants released by moving vehicles are more harmful than pollutants from other sources (Kumar 2020).

4.4 Biomonitoring of HMs

In a quality control program, biological monitoring entails the routine use of living things to gather quantitative data on environmental changes that are frequently brought on by human activity. Because they cover a wider range of space and time than data from chemical or physical detectors, biological responses can be thought of as more accurate representations of environmental conditions. Additionally, they enable estimation of the concentrations of pollutants and, more crucially, their effects on biological receptors. Plants at every level can serve as bioindicators, biomonitoring devices, and bio-accumulators. Generally, a range of scale levels, from the molecular to the population or community rank, are used to evaluate the reactions of vegetative structures. Even however, a certain quantity of pollution may not be enough to cause the vegetative structures to respond, even while it has a significant impact on plants' total ability to reproduce. Pollen does appear to be a particularly good indication of environmental toxins. In reality, several genes regulate pollen quality, which can be assessed using a variety of techniques. Virtually any tiny loss in the genome would result in pollen abortion. Pollen, the male haploid generation of higher plants, is used less frequently than other methods (Calzoni et al. 2007).

Forests and other forms of plants have been shown to be affected by heavy metal accumulating. As biological trace element monitors, terrestrial and aquatic plant species are being employed more and more. A species that is widely distributed, simple to recognize, and has a high tolerance to environmental toxins makes for an effective biological monitor. Metals accumulating in the monitor body need to be an accurate representation of the outside world. Often cited in environmental studies, the dandelion (Taraxacum officinale, Weber) is a common plant that may be found in many environments. The assessment of SO_2, polycyclic aromatic hydrocarbons, and heavy metal contamination is one of its applications. In earlier research, there has been a correlation between the quantity of many metals in Taraxacum officinale leaves and nearby environmental contamination (Kleckerová and Dočekalová 2014).

4.5 Potential Ecological Risks

There are many different human activities that lead to the degradation of the environment that is brought on by pollution. Unwanted environmental changes brought on by pollution have negative effects on people, animals, and plants. The kind and quantity of pollutants affect how badly they affect the environment and the health of people (Asiminicesei et al. 2020). The effects of solid wastes of anthropogenic origin on the environment are becoming more widely known (Gavrilescu 2010; Alharbi et al. 2018).

The majority of the time, dumpsites are abandoned agricultural fields, and the garbage that is irresponsibly dumped there has a terrible impact on the soil, water, and air. The sort of vegetation that may grow in a particular soil is often influenced by its physicochemical characteristics. For instance, soil structure and acidity have an impact on how well plants absorb and store minerals (Ekere et al. 2020; Gavrilescu et al. 2015). Furthermore, the concentrations and types of heavy metals in soil and consequently in crops around dumpsites are influenced by the waste's types, run-off, topography, and level of scavenging (Fiedler 2008).

Usually, it comprises poisonous pesticides and substances that may physically harm pests, such as silicate, borate, and sulfur compounds, that are recovered from mines and then turned into powders. Some of these pesticides are extremely dangerous since they include metals including arsenic, copper, lead, and tin salts. As persistent inorganic pollutants, heavy metals are nonbiodegradable, can accumulate in soils and biological compartments and move through the food chain, then affecting the normal functions of the human body (Živković et al. 2012; Alkherraz et al. 2013; Asiminicesei et al. 2020). The dangers posed by the presence of heavy metals in the environment are covered in great detail in the literature on various contaminants.

Heavy metal contamination of agricultural soil may pose health risk to humans via the food chain and can also lead to reduction in food quality (Gavrilescu 2010). Heavy metals have a stronger propensity to accumulate in crops and vegetables growing in polluted soil. Consumption of vegetable has increased in recent years due to its health benefits and uptake of metals by vegetables is a major pathway for soil metals to enter the food chain and bio-accumulates leading to human health risk (Hlihor et al. 2015). The chemical make-up of the waste, its physical attributes, the type of vegetables grown, and the pace of consumption all affect the health hazards. Through inhalation, food, and manual handling, heavy metals from polluted soils make their way into plants and then into human tissues. The metals can attach to crucial biological elements including structural proteins, enzymes, and nucleic acids and prevent them from working properly. Long-term exposure to heavy metals can have negative effects on the circulatory system, central nervous system, and peripheral nervous system. To guarantee that the acceptable values are not exceeded, it is necessary to regularly check the amount of heavy metal bioaccumulation in edible crops (Ekere et al. 2020).

Heavy metals are significant environmental pollutants, and their toxicity significantly affects ecological, evolutionary, nutritional, and environmental systems (Ibrahim et al. 2017). Long-term irrigated regions with treated or untreated wastewater, places with high automobile traffic, tailings dumps, or other forms of waste deposits that may include metals are the origins of heavy metal contamination of medicinal plants (Asiminicesei et al. 2020). Soils in various parts of the world are slightly to moderately contaminated with toxic heavy metals, such as Cd, Cu, Zn, Ni, Co, Cr, Pb, and As, as a consequence of long-term use of phosphate fertilizers, soil amendment with sludge from wastewater treatment plants, traffic, the presence of industrial waste, and inadequate irrigation practices (Pavel et al. 2010; Yadav 2010). The most frequent hazardous metals are lead (Pb), chromium (Cr), arsenic (As), zinc (Zn), cadmium (Cd), copper (Cu), mercury (Hg), and nickel. Heavy metals are persistent toxic compounds that may also cause significant soil and water contamination (Ni). The majority of metals may stay in soils for a very long time after being introduced because they cannot be broken down by microbial or chemical activity. Toxic metals in the soil can severely inhibit the biodegradation processes of some organic pollutants (Wuana and Okieimen 2011; Manu et al. 2018; Asiminicesei et al. 2020). Being stationary creatures, plants continually interact with heavy metal-polluted soil. To combat these circumstances, plants have evolved sophisticated defense mechanisms that use a wide range of chemical compounds. Studies have shown that metals that are soluble in soil solutions or that are solubilized by exudates from plant roots are therefore bioavailable for uptake by plants (Asiminicesei et al. 2020). Although plants need a number of metals for growth, development and maintenance, they can become toxic if they are in excessive amounts, but the ability of plants to accumulate essential metals makes it possible to absorb other metals, whose presence affects the plant (Pavel et al. 2010; Ibrahim et al. 2017; Sobariu et al. 2017). Some plants are bred to take heavy metals from the soil and purify the environment because they can withstand high concentrations of the pollutants. In contrast, agricultural crops, industrial plants, or other useful veggies meant for human food and use make up the bulk of plants grown on polluted soils. Therefore, eating fruits and vegetables from plants that were grown in soil with high levels of heavy metals might have a negative impact on one's health. For example, some studies estimate that about half of the lead in the human body comes from food, which in a proportion of approximately 50% originate from plants (Intawongse and Dean 2006; Asiminicesei et al. 2020.

Despite being used as natural remedies, foods, and sources of nourishment, medicinal plants can nevertheless be polluted with various organic or inorganic toxins. For instance, heavy metals and other persistent inorganic pollutants are regarded as a group of high concern due to their detrimental effects on both the environment and human health. Contrary to many organic contaminants, which biodegrade into carbon dioxide and water, heavy metals may accumulate in the environment and, as previously indicated, have a multitude of detrimental consequences on both the ecosystem and human health. Although some medicinal plants are harvested from the wild, their cultivation for commercial purposes has grown directly in proportion to the requirement (Ibrahim et al. 2017; Asiminicesei et al.

2020). Heavy metal deposition in plant tissues can be accelerated by pesticide usage and poor storage conditions for medicinal plants after harvest.

Some heavy metals are essential to the organism, such as Cu, Mn, and Zn, which serve as part of proteins and enzymes in human body with the function of enhancing interaction and deactivating enzyme activity (Gavrilescu 2010; Gavrilescu et al. 2015). However, after building up in specific bodily organs, they may also result in chronic poisoning. Therefore, consuming too many necessary heavy metals is bad for the human body. Other heavy metals, meanwhile, are harmful to the organism and useless. Herbal remedies may include significant levels of dangerous metals that are scarcely biodegradable, such as, Cd, or Pb. These metals may be absorbed by the roots from the soil, deposited in the air through wet or dry deposition, or polluted during processing. Thus, they are enriched in human body by food chain for thousand folds, which is equal to biological amplification (Ncube et al. 2012; Ekere et al. 2020). In order to ensure the secure use of herbs, the World Health Organization (WHO) has defined maximum concentration limits for hazardous substances (1996). In order to protect human health, rigorous restrictions have been placed on the amount of heavy metals that may be found in soil and medications.

Additionally, vehicle pollution has an impact on how plants function physiologically. By absorbing heavy metals, which impact plants' anatomical, physiological, and reproductive characteristics, the pH of soil is progressively changed. Plant reproductive organs are negatively impacted by car emissions. Roadside vegetation's germination and seedling development are impacted by heavy metal pollution. Overly toxic Pb causes a reduction in the germination rate of seedlings. Controlling heavy metal contamination is essential for ecosystem recovery and restoration. To counteract the negative impacts of heavy metal contamination, several remediation solutions have been created. A simple and effective method for reducing the pollution caused by heavy metals is phytoremediation. Various plant species with a high potential to absorb heavy metals can be employed in phytoremediation (Altaf et al. 2021).

5 Assessment of Ecological Risk by HMs

Ecological risk assessment (EcoRA) entails evaluating the dangers that, in principle, exist for all living things in the many ecosystems that make up the environment due to the presence of chemicals introduced into the environment by humans. Technical assistance for management choices such as whether to allow a discharge into a body of water, whether or how much to clean up a spill, or what harvest restrictions to set for a natural resource is provided by ecological risk assessments. Similar to other risk assessment techniques, ecological risk assessment focuses on determining the kind, scope, and probability of unfavorable outcomes of situations or activities. It stands out for being concerned about ecological repercussions (effects on nonhuman organisms, populations, and communities). It is a procedure that includes issue conceptualization, exposure analysis, effect analysis, and risk characterization. The scope and procedures of the assessment process are decided upon during problem formulation, which also yields a conceptual model of the system under evaluation.

Numerous creatures, each with a different sensitivity to chemicals, and different groups with different exposure situations, such as free swimmers and sediment dwellers, are dealt with by EcoRA. The accepted approach is to test chosen members of important taxonomic groups and use them as substitutes for the entire system because it is impossible to gather toxicity data on all the creatures in an ecosystem. This approach raises concerns since it could not safeguard the most vulnerable species that are exposed to the environment. Widespread harm to animals and ecosystems can occur if the actions of a chemical on a possible receptor are not identified. An example of this is the harm done to oysters and dog whelks as a result of using antifouling paints that contain tributyltin (Suter and Norton 2019). Although hazard identification and dose–response assessment is consolidated into a single stage, the process still includes the three elements required for health risk assessment. Effects assessment:

1. A predicted no effect concentration (PNEC), generated from ecotoxicity data and the application of assessment variables, is estimated as part of the effects assessment process. This comprises identifying the danger based on its physico-chemical characteristics, ecotoxicity, and planned use.
2. Calculating a predicted environmental concentration (PEC) is part of the exposure assessment process. Monitoring data, plausible worst-case scenarios, and predictive modeling approaches are used to determine this. It is a difficult undertaking that should take release, degradation, transport, and destiny processes into account.
3. The PEC/PNEC ratio is a quotient that is calculated as part of the risk characterization process. The material is thought to pose no damage to the environment in a particular situation if the ratio is less than 1.

Various indexes that are used for the assessment of ecological risks are discussed below. These indexes are important to show how the value changes with heavy metals concentrations in environment.

5.1 *Various Indexes*

5.1.1 Air Pollution Tolerance Index (APTI)

The ability of plants to withstand air pollution is described by the air pollution tolerance index (APTI). It is one of the crucial factors that might be taken into account while choosing the species of plants for traffic barriers. The ability to classify plants according to their levels of sensitivity or tolerance to air pollution is a crucial tool. Four biochemical factors have been used to define plant APTI: total chlorophyll, relative water content (RWC), ascorbic acid, and pH of leaf extract. A single parameter's change as a result of pollution may not accurately represent the situation. Consequently, four biochemical variables are taken into account to determine an empirical value for plants' APTI (Nadgórska-Socha et al. 2017).

One of the frequent consequences of air pollution that inhibits photosynthesis is the progressive removal of chlorophyll and the yellowing of leaves. An essential electron donor in photosynthesis is ascorbic acid. pH is a key factor in photosynthetic activity. These metabolic factors are interdependent and most affected by air pollution. Low APTI plant species can be utilized as bio-indicators in low pollution regions based on APTI values, whereas high APTI plant species can be used to reduce air pollution in highly contaminated areas (Shrestha et al. 2021).

ATPI can be calculated by below formula:

$$\mathrm{ATPI} = [A(T + P) + (R)]/10$$

where A = Ascorbic acid content (mg g^{-1}) (dry weight), T = total chlorophyll content (mg g^{-1}), and P = leaf extract pH, and R = RWC (%).

5.1.2 Bio-Contamination Factor (BCF)

The heavy metals and trace elements are transported from the soil mixture to the plants according to the bio-concentration factor (BCF). Any metal's BCF may be determined by dividing the quantity of heavy metal in a plant sample by the same amount in the matching soil sample. The formula below was used to determine the BCF:

$$\mathrm{BCF} = C_{\mathrm{Plant}}/C_{\mathrm{Soil}}$$

where C_{Soil} is the amount of heavy metals in the associated soil, and C_{Plant} is the amount of heavy metals in the plant (dry weight). If BCF is less than 1, the plant merely absorbed the heavy metal; it did not accumulate. If BCF is more than 1, the plant both absorbed and accumulated the heavy metal (Kaur et al. 2022).

5.1.3 Translocation Factor (TF)

Translocation factor (TF) is the capacity for translocation from the plant's root to its aboveground components (the stem and leaves), which can calculate using the formula:

$$\mathrm{TF} = C_{\mathrm{stem\ or\ leaves}}/C_{\mathrm{root\ or\ stem}}$$

Metal concentrations in stems and leaves are expressed as $C_{\mathrm{stem\ or\ leaves}}$ and in roots or stems as $C_{\mathrm{root\ or\ stem}}$ (Sulaiman 2018).

5.1.4 Enrichment Factor (EF)

According to Chester et al. (1999), the enrichment factors can be used to determine how enriched or depleted heavy metals are in relation to a certain source (EF) (Chester et al. 1999). Additionally, it aids in separating anthropogenic origins from crustal sources of heavy metals in road dust. Estimates were made of the EFs of heavy metals in dust in relation to their crustal compositions. It is calculated using the equation shown below:

$$\mathrm{EF} = (\mathrm{HM}/x)_{\mathrm{dust}}/(\mathrm{HM}/x)_{\mathrm{crust}}$$

where $(\mathrm{HM}/x)_{\mathrm{dust}}$ is the concentration ratio of the heavy metal (HM) to x in dust samples, $(\mathrm{HM}/x)_{\mathrm{crust}}$ is their average ratio in the upper continental crust, and x is the amount of most abundant metal in results.

There were five contamination categories: where $\mathrm{EF} < 2$ represents the depletion of metal enrichment; $2 \leq \mathrm{EF} < 5$ moderate enrichment; $5 \leq \mathrm{EF} < 20$ significant enrichment; $20 \leq \mathrm{EF} < 40$ very high enrichment; and $\mathrm{EF} > 40$ extremely high enrichment. In general, the EF value of$>$10, enriched element, is an indicator of anthropogenic source, whereas EF value of$<$10 indicates crustal element (Al-Taani et al. 2019; Ekere et al. 2020).

5.1.5 Contamination Factor (CF)

The ratio of the metals in the samples to the background value of the relevant metal was used to calculate the contamination factor (CF).

Pekey et al. (2004) divided the contamination factor into four categories: low ($\mathrm{CF} < 1$), moderate ($1 \leq \mathrm{CF} \leq 3$), considerable ($3 \leq \mathrm{CF} \leq 6$), and very high ($\mathrm{CF} > 6$) (Pekey et al. 2004; Sulaiman 2018).

5.1.6 Index of Geo-Accumulation (Igeo)

Index of geo-accumulation (Igeo) values varied from 0 to 5, where 0 is uncontaminated soil and 5 is very contaminated soil, and were determined using an equation initially presented by Muller in 1997 to assess the degree of soil contamination by heavy metals (Singh et al. 1997).

$$\mathrm{Igeo} = \mathrm{Log}_2\ (\mathrm{Cm}/1.5^{*}\mathrm{Bm}),$$

where 1.5 is a factor used to account for any fluctuations in the background data caused by various lithogenic influences, Cm is the concentration of the studied metal in the soil sample, and Bm is the background level of the same metal (Kaur et al. 2022).

5.1.7 Pollution Load Index (PLI)

Pollutant load index (PLI) was introduced by Tomlinson. The following formula was used to determine the pollutant load index (PLI) for metals:

$$\mathrm{PLI} = (\mathrm{CF}_1 \times \mathrm{CF}_2 \ldots \times \mathrm{CF}_n)^{1/n}$$

where n is the number of heavy metals and CF is the computed contamination factor. The PLI offers straightforward but comparable methods for rating the quality of a study site. Each metal's PLI was divided into three categories: baseline level ($\mathrm{PLI} = 1$), not polluted ($\mathrm{PLI} < 1$), and polluted ($\mathrm{PLI} > 1$) (Al-Taani et al. 2019).

5.1.8 Potential Ecological Risk Index (PERI)

The potential ecological risk index (PERI), which was established by Hakanson (1980), was used to assess the impact of various metal contaminations in dust using following formula:

$$\mathrm{PERI} = \Sigma Ei,$$

$$Ei = Ti \times \mathrm{CF}$$

$$\mathrm{CFi} = C_m / B_m$$

where Ti is the toxic reaction factor for a particular heavy metal I and Ei is the single ecological risk index. According to Hakanson (1980) and Xu et al. (2008), the Ti values for Hg, Cd, As, Ni, Pb, Cr, Mn, and Zn are 40, 30, 10, 5, 2, 1, and 1, respectively [52, 8]. Ei can be classifed as either low risk ($Ei < 40$), moderate ($40 \leq Ei < 80$), considerable risk ($80 \leq Ei < 160$), high ($160 \leq Ei < 320$), or signifcantly high risk ($Ei \geq 320$). The PERI consists of four grades: <150 low, $150 \leq \mathrm{PERI} < 300$ moderate, $300 \leq \mathrm{PERI} < 600$ considerable, and ≥ 600 very high (Hakanson 1980; Xu et al. 2008).

6 Conclusions

Usually focusing on airborne pollution, tailpipe (silencer) emission restrictions address concerns about the impact of motor vehicle emissions on human health. But automobiles also produce heavy metal pollutants via non-tailpipe sources (e.g., brake wear and tire particulates). Long after they have descended from the stratosphere, the metal contaminants created pose risks to ecosystem and human health, mainly via contaminating soils and plants. Through photosynthesis, plants are known to bind carbon dioxide (CO2), making them a significant resource in efforts to reduce air pollution. In many nations, common plant species line certain roads and national highways; however, these plants, such as *Azadirachta indica*, *Bougainvillea spectabilis*, *Cassia fistula*, *Ficus religiosa*, and *Polyalthia iongifolia*, are harmed by heavy metal air pollution (Ni, Pb, Cr, Zn, Cu, and Cd). In this way, the varieties of trees that line the sides of roadways and in close proximity to them serve as a buffer against vehicle emissions.

6.1 Future Aspects

For the purpose of reducing air pollution, urban and peri-urban vegetation is being considered. The requirement for developing a green area is suitable plants that are effective at adsorbing and absorbing air pollutants. To evaluate how well roadside plantings can withstand urban roadside air pollution brought on by high traffic volumes, physiological- and biophysical-based parameters are combined. The assumption was made that *Azadirachta indica*, *Bougainvillea spectabilis*, *Cassia fistula*, *Ficus religiosa*, and *Polyalthia iongifolia,* etc. are a species that is tolerant of vehicle emissions and may be utilized to create roadside plantings and other green areas in an urban setting.

The field of ecological risk assessment is still very much in its infancy, and there are still many issues that need to be solved, including:

- Determining the impacts on populations and communities.
- Selection of end points.
- Selection of indicative species.
- The decision about field, lab, mesocosm, and microcosm experiments.
- The inclusion of ecological recovery and resilience variables.

References

Al-Chalabi AS, Hawker D (2000) Distribution of vehicular lead in roadside soils of major roads of Brisbane, Australia. Water Air Soil Pollut 118(3):299–310

Alharbi OML, Khattab RA, Ali I (2018) Health and environmental effects of persistent organic pollutants. J Mol Liq 263:442–453

Alkherraz AM, Amer AM, Mlitan AM (2013) Determination of some heavy metals in four medicinal plants. World Acad Sci Eng Technol 78:1568–1570

Al-Taani AA, Nazzal Y, Howari FM (2019) Assessment of heavy metals in roadside dust along the Abu Dhabi–Al Ain National Highway, UAE. Environ Earth Sci 78(14)

Altaf R, Altaf S, Hussain M, Shah RU, Ullah R, Ullah MI, Rauf A, Ansari MJ, Alharbi SA, Alfarraj S, Datta R (2021) Heavy metal accumulation by roadside vegetation and implications for pollution control. PloS One 16

Amaya E, Gil F, Freire C, Olmedo P, Fernandez-Rodriguez M, Fernandez MF, Olea N (2013) Placental concentrations of heavy metals in a mother–child cohort. Environ Res 120:63–70

Ashraf MA, Maah MJ, Yusoff I (2011) Heavy metals accumulation in plants growing in ex tin mining catchment. Int J Environ Sci Technol 8(2):401–416

Asiminicesei DM, Vasilachi IC, Gavrilescu M (2020) Heavy metal contamination of medicinal plants and potential implications on human health. Rev Chim 71(7):16–36
Avnish C (2008) Impact of dust pollution on photosynthetic pigments of some selected trees grown at nearby of stone-crushers. Environ Conserv J 9(3):11–13
Babin-Fenske J, Anand M (2011) Patterns of insect communities along a stress gradient following decommissioning of a Cu–Ni smelter. Environ Pollut 159(10):3036–3043
Bakirdere S, Yaman M (2008) Determination of lead, cadmium and copper in roadside soil and plants in Elazig, Turkey. Environ Monit Assess 136(1):401–410
Bernhardt-Römermann M, Kudernatsch T, Pfadenhauer J, Kirchner M, Jakobi G, Fischer A (2007) Long-term effects of nitrogen deposition on vegetation in a deciduous forest near Munich, Germany. Appl Veg Sci 10(3):399–406
Bobbink R, Hornung M, Roelofs JGM (1998) The effects of air-borne nitrogen pollutants on species diversity in natural and semi-natural European vegetation. J Ecol 86(5):717–738
Bohemen HD, van Janssen Van De Laak WH (2003) The influence of road infrastructure and traffic on soil, water, and air quality. Environ Manag 31(1):50–68
Calzoni GL, Antognoni F, Pari E, Fonti P, Gnes A, Speranza A (2007) Active biomonitoring of heavy metal pollution using Rosa rugosa plants. Environ Pollut 149(2):239–245
Chauhan A (2010) Photosynthetic pigment changes in some selected trees induced by automobile exhaust in Dehradun, Uttarakhand. N Y Sci J 3(2):45–51
Chester R, Nimmo M, Preston MR (1999) The trace metal chemistry of atmospheric dry deposition samples collected at Cap Ferrat: a coastal site in the Western Mediterranean. Mar Chem 68(1–2): 15–30
Dzwonko Z (2001) Assessment of light and soil conditions in ancient and recent woodlands by Ellenberg indicator values. J Appl Ecol 38(5):942–951
Ekere NR, Ugbor MCJ, Ihedioha JN, Ukwueze NN, Abugu HO (2020) Ecological and potential health risk assessment of heavy metals in soils and food crops grown in abandoned urban open waste dumpsite. J Environ Health Sci Eng 18(2):711–721
Elik A (2003) Heavy metal accumulation in street dust samples in Sivas. Commun Soil Sci Plant Anal 34(1–2):145–156
Fakayode SO, Olu-Owolabi BI (2003) Heavy metal contamination of roadside topsoil in Osogbo, Nigeria: its relationship to traffic density and proximity to highways. Environ Geol 44(2): 150–157
Fang T, Jiang T, Yang K, Li J, Liang Y, Zhao X, Gao N, Li H, Lu W, Cui K (2021) Biomonitoring of heavy metal contamination with roadside trees from metropolitan area of Hefei, China. Environ Monit Assess 193(3)
Feng J, Zhao J, Bian X, Zhang W (2012) Spatial distribution and controlling factors of heavy metals contents in paddy soil and crop grains of rice–wheat cropping system along highway in East China. Environ Geochem Health 34(5):605–614
Fiedler H (2008) Stockholm convention on POPs: obligations and implementation. The fate of persistent organic pollutants in the environment. Springer, pp 3–12
Gall JE, Boyd RS, Rajakaruna N (2015) Transfer of heavy metals through terrestrial food webs: a review. Environ Monit Assess 187(4):1–21
Gavrilescu M (2010) Biosorption in environmental remediation. Bioremediation technology. Springer, pp 35–99
Gavrilescu M, Demnerová K, Aamand J, Agathos S, Fava F (2015) Emerging pollutants in the environment: present and future challenges in biomonitoring, ecological risks and bioremediation. N Biotechnol 32(1):147–156
Gope M, Masto RE, George J, Balachandran S (2018) Tracing source, distribution and health risk of potentially harmful elements (PHEs) in street dust of Durgapur, India. Ecotoxicol Environ Saf 154:280–293
Hakanson L (1980) An ecological risk index for aquatic pollution control.a sedimentological approach. Water Res 14(8):975–1001

Harrison RM, Laxen DPH (1977) A comparative study of methods for the analysis of total lead in soils. Water Air Soil Pollut 8(4):387–392

Hlihor RM, Diaconu M, Leon F, Curteanu S, Tavares T, Gavrilescu M (2015) Experimental analysis and mathematical prediction of Cd (II) removal by biosorption using support vector machines and genetic algorithms. N Biotechnol 32(3):358–368

Ibrahim MH, Chee Kong Y, Mohd Zain NA (2017) Effect of cadmium and copper exposure on growth, secondary metabolites and antioxidant activity in the medicinal plant Sambung Nyawa (Gynura procumbens (Lour.) Merr). Molecules 22(10):1623

Intawongse M, Dean JR (2006) Uptake of heavy metals by vegetable plants grown on contaminated soil and their bioavailability in the human gastrointestinal tract. Food Addit Contam 23(1): 36–48

Jia L, Wang W, Li Y, Yang L (2010) Heavy metals in soil and crops of an intensively farmed area: a case study in Yucheng City, Shandong Province, China. Int J Environ Res Public Health 7(2): 395–412

Kaur M, Kaur R, Singh N, Saini S, Katnoria JK, Nagpal AK (2022) Assessment of genotoxic and tumorigenic potential and heavy metal contamination in roadside soil and plants of Amritsar (Punjab), India. Environ Earth Sci 81(4)

Khalid N, Hussain M, Young HS, Boyce B, Aqeel M, Noman A (2018) Effects of road proximity on heavy metal concentrations in soils and common roadside plants in Southern California. Environ Sci Pollut Res 25(35):35257–35265

Khalifa M, Gad A (2018) Assessment of heavy metals contamination in agricultural soil of southwestern Nile Delta, Egypt. Soil Sediment Contam Int J 27(7):619–642

Kleckerová A, Dočekalová H (2014) Dandelion plants as a biomonitor of urban area contamination by heavy metals. Int J Environ Res 8(1):157–164

Kováts N, Hubai K, Diósi D, Sainnokhoi TA, Hoffer A, Tóth Á, Teke G (2021) Sensitivity of typical European roadside plants to atmospheric particulate matter. Ecol Indic 124

Kumar A (2020) Adaptation and mitigation potential of roadside trees with bio-extraction of heavy metals under vehicular emissions and their impact on physiological traits during seasonal regimes

Lee MA, Davies L, Power SA (2012) Effects of roads on adjacent plant community composition and ecosystem function: an example from three calcareous ecosystems. Environ Pollut 163: 273–280

Li X, Poon C, Liu PS (2001) Heavy metal contamination of urban soils and street dusts in Hong Kong. Appl Geochem 16(11–12):1361–1368

Manu M, Onete M, Băncilă RI (2018) The effect of heavy metals on mite communities (Acari: Gamasina) from urban parks-Bucharest, Romania. Environ Eng Manag J 17(9)

Miranda M, Benedito JL, Blanco-Penedo I, López-Lamas C, Merino A, López-Alonso M (2009) Metal accumulation in cattle raised in a serpentine-soil area: relationship between metal concentrations in soil, forage and animal tissues. J Trace Elem Med Biol 23(3):231–238

Mukherjee A, Agrawal M (2018) Use of GLM approach to assess the responses of tropical trees to urban air pollution in relation to leaf functional traits and tree characteristics. Ecotoxicol Environ Saf 152:42–54

Nadgórska-Socha A, Kandziora-Ciupa M, Trzęsicki M, Barczyk G (2017) Air pollution tolerance index and heavy metal bioaccumulation in selected plant species from urban biotopes. Chemosphere 183:471–482

Naser HM, Sultana S, Gomes R, Noor S (2012) Heavy metal pollution of soil and vegetable grown near roadside at Gazipur. Bangladesh J Agric Res 37(1):9–17

Ncube B, Finnie JF, van Staden J (2012) Quality from the field: the impact of environmental factors as quality determinants in medicinal plants. S Afr J Bot 82:11–20

Pagotto C, Remy N, Legret M, le Cloirec P (2001) Heavy metal pollution of road dust and roadside soil near a major rural highway. Environ Technol 22(3):307–319

Pan J, Plant JA, Voulvoulis N, Oates CJ, Ihlenfeld C (2010) Cadmium levels in Europe: implications for human health. Environ Geochem Health 32(1):1–12

Pavel VL, Bulgariu D, Bulgariu L, Hlihor RM, Gavrilescu M (2010) Analysis of factors determining the behaviour of chromium in some Romanian soils. Environ Eng Manag J 9(1):89–94
Pekey H, Karakaş D, Ayberk S, Tolun L, Bakoğlu M (2004) Ecological risk assessment using trace elements from surface sediments of İzmit Bay (northeastern Marmara Sea) Turkey. Mar Pollut Bull 48(9–10):946–953
Princewill-Ogbonna IL, Ogbonna PC (2011) Heavy metal content in soil and medicinal plants in high traffic urban area. Pak J Nutr 10(7):618–624
Sadia HE, Jeba F, Uddin MZ, Salam A (2019) Sensitivity study of plant species due to traffic emitted air pollutants (NO2 and PM2.5) during different seasons in Dhaka, Bangladesh. SN Appl Sci 1(11)
Sánchez-Chardi A (2016) Biomonitoring potential of five sympatric tillandsia species for evaluating urban metal pollution (Cd, Hg and Pb). Atmos Environ 131:352–359
Shahid M, Dumat C, Khalid S, Schreck E, Xiong T, Khan Niazi N, Foliar A (2016) Foliar heavy metal uptake, toxicity and detoxification in plants: a comparison of foliar and root metal uptake. J Hazard Mater:325
Shrestha S, Baral B, Dhital NB, Yang HH (2021) Assessing air pollution tolerance of plant species in vegetation traffic barriers in Kathmandu Valley, Nepal. Sustain Environ Res 31(1)
Sinegani AAS (2007) Temporal and spatial variability of lead levels in Salsola kali near Razan-Hamadan highway
Singh M, Ansari AA, Müller G, Singh IB (1997) Heavy metals in freshly deposited sediments of the Gomati River (a tributary of the Ganga River): effects of human activities. Environ Geol 29: 246–252
Singh H, Yadav M, Kumar N, Kumar A, Kumar M (2020) Assessing adaptation and mitigation potential of roadside trees under the influence of vehicular emissions: a case study of Grevillea robusta and Mangifera indica planted in an urban city of India. PloS One 15(1):e0227380
Sobariu DL, Fertu DIT, Diaconu M, Pavel LV, Hlihor R-M, Drăgoi EN, Curteanu S, Lenz M, Corvini PF-X, Gavrilescu M (2017) Rhizobacteria and plant symbiosis in heavy metal uptake and its implications for soil bioremediation. N Biotechnol 39:125–134
Stankovic S, Kalaba P, Stankovic AR (2014) Biota as toxic metal indicators. Environ Chem Lett 12(1):63–84
Sulaiman FR (2018) Heavy metals accumulation in suburban roadside plants of a tropical area (Jengka, Malaysia)
Suter GW, Norton SB (2019) Ecological risk assessment. Encyclopedia Ecol:402–406
van der Fels-Klerx HJ, Romkens P, Franz E, van Raamsdonk LWD (2011) Modeling cadmium in the feed chain and cattle organs. Biotechnol Agron Soc Environ 15(Special Issue 1):53–59
Wuana RA, Okieimen FE (2011) Heavy metals in contaminated soils: a review of sources, chemistry, risks and best available strategies for remediation. Int Sch Res Notices
Xu L, Yu Z, Lee H-M, Wolff MS, Golub LM, Sorsa T, Kuula H (2008) Characteristics of collagenase-2 from gingival crevicular fluid and peri-implant sulcular fluid in periodontitis and peri-implantitis patients: pilot study. Acta Odontol Scand 66(4):219–224
Yadav SK (2010) Heavy metals toxicity in plants: an overview on the role of glutathione and phytochelatins in heavy metal stress tolerance of plants. S Afr J Bot 76(2):167–179
Živković J, Ražić S, Arsenijević J, Maksimović Z (2012) Heavy metal contents in Veronica species and soil from mountainous areas in Serbia. J Serb Chem Soc 77(7):959–970

Sustainable Utilization of Anthropogenic Coal Fly Ash Through Mechanical and Chemical Activation

Dilip Kumar Rajak, Swapan Suman, Chandan Guria, and Ganesh Kumar

Abstract Anthropogenic coal fly ash (CFA) is a pozzolanic material comprises silico-alumina, making it an excellent secondary raw material with a range of applications. It is utilized for wastewater treatment, extraction of valuable minerals, and the production of ceramics, cement, concrete, building materials, composites, paints, and plastic materials. The crystalline and amorphous phases of CFA contain metals and metalloid oxides, with the amorphous portion playing a significant role in chemical reactions. However, the direct use of coal fly ash poses challenges due enriched potentially toxic trace elements. Nevertheless, the upstream extraction of valuable minerals and efficient downstream applications of CFA can be improved through mechanical and/or chemical activation. Industrial and laboratory-scale purification and modification of coal fly ash by size reduction, surface modification, and functionalization are discussed in this chapter.

Keywords Anthropogenic · Downstream · Mechanical and chemical activation · Upstream

---

D. K. Rajak (✉)
Department of Chemistry, University of the Free State, Bloemfontein, South Africa

Department of Applied Sciences and Chemical Engineering, The Institute of Engineering, Pulchowk Campus, Tribhuvan University, Kathmandu, Nepal

S. Suman
Department of Mechanical Engineering, MIET, Meerut, Uttar Pradesh, India

C. Guria
Department of Petroleum Engineering, Indian Institute of Technology (Indian School of Mines), Dhanbad, Jharkhand, India

G. Kumar
Department of Ocean Engineering, Indian Institute of Technology Madras, Chennai, India

P. Pathak et al. (eds.), *Anthropogenic Environmental Hazards*,
https://doi.org/10.1007/978-3-031-41013-0_7

1 Introduction

Coal fly ash (CFA) is the thermal power plant waste produced from burning pulverized coal and is a hazard to the environment. CFA particle contains potentially toxic trace elements. CFA is generally used for landfill, causing serious environmental concerns. CFA is made up of abundant anthropogenic materials, which are composed of quartz, mullite, hematite, calcite, and rutile. These elements pose severe effects on the environment, especially in soil, air, and human health (Carlson and Adriano 1993). Water-soluble constituents of CFA are added to soil which increase its salinity and pH and decrease the concentration of water-soluble phosphorous, ultimately affecting soil fertility (Page et al. 1979). Depending on the chemical properties and toxicity, CFA particles present in the air are a hazard to the workers when they inhale (Borm 1997). Numerous researchers have dedicated their efforts to exploring the utilization of CFA as a means to mitigate its detrimental environmental consequences. These investigations involve either the direct application of CFA or the creation of novel products through mechanical activation and/or chemical activation techniques.

CFA is a pozzolanic powdery material rich in silico-aluminate, abundant in anthropogenic materials composed of crystalline and amorphous phases (Vassilev et al. 2003). CFA falls as the fifth largest raw material (Ahmaruzzaman 2010), which is currently used in the concrete making as a cement additive to improve their workability and durability (Bilodeau and Malhotra 2000). Around 44.26% of world CFA is used in the cement sector, 13% in mine filling, 12.88% in embankments and roads making, 10.77% in low-lying area reclamation, 1.93% in soil fertility improvement, and 11.72% in the making of bricks and tiles in the recent years (Singh et al. 2016). Despite being employed in various sectors, the utilization of CFA remains limited, with only 55.69% of the total CFA being effectively utilized. Moreover, it finds applications in diverse areas such as valuable mineral extraction and separation, wastewater treatment, geopolymer precursor and zeolite production, as well as in composite, concrete, and paint materials (Ahmaruzzaman 2010; Juenger et al. 2011; Srivastava et al. 2023).

Raw CFA complicates the hydration reaction in cement, resulting in a longer setting time and a delay in the development of early-age strength in concrete. However, by carefully modifying factors such as particle size distribution, surface smoothness, shape morphology, fineness, compressive strength, bulk density, porosity, permeability, pozzolanic activity, and the radical composition of both the crystalline and amorphous phases of CFA, its utilization can be enhanced (Bentz 2010; Bentz and Ferraris 2010; Rajak et al. 2017). These properties can be achieved through the mechanical activation (size reduction) of CFA using a high-energy mill. Ultimately mechanically activated CFA contains more finer fractions with an increase in glass content (amorphous siliceous spherical particulates) and also increases the reaction rate and decreases the reaction time. The mechanically activated CFA has relatively high reactivity in alkaline solution and promotes direct transformation into zeolite and geopolymer materials (Ahmaruzzaman 2010; Kumar and Kumar 2011; Rajak et al. 2017; Blicharz et al. 2022).

Chemical activation, which includes surface modification and/or functionalization, also improves the utilization of CFA as it provides twofold benefits (a) increased reactive site by spoiling smooth and close surface structure of CFA glasses and (b) formation of three-dimensional inorganic polymer like material, which can be easily subjected to beneficiary reactions (Provis and Van Deventer 2013; Rajak et al. 2016; Gautam et al. 2018; Gollakota et al. 2019).

In the past two decades, alkali-treated CFA has gained significant popularity as a means to synthesize cost effective and environmentally friendly cement-like materials (Nidheesh and Kumar 2019; Srivastava et al. 2023). Notably, the crystalline and amorphous phases of CFA predominantly consist of mullite (Al_2O_3) and quartz (SiO_2), the proportions of which vary depending on the type of coal used in power plants. Alkali reacts with amorphous SiO_2 of CFA in an aqueous solution, whereas the crystalline part remains unaffected (Pietersen 1990; Rowles and O'Connor 2009; Rajak et al. 2016).

In the present study, a brief study is done on the characterization of CFA, which includes morphological analysis, physical (area to mass ratio, gravity, and particle size distribution), and chemical properties (radical compositions) of CFA, which helps to classify the CFA. The characterization of CFA also provides a full-proof path for mechanical and chemical activation. In the mechanical activation, the determination of breakage parameters for CFA, such as the breakage distribution parameter ($B_{i,j}$), and specific rate of breakage (S_i) using a planetary ball mill are discussed, as these kinetic parameters give an idea of how much mass fractions with time the CFA are grinded. In addition to mechanical activation, the present study also investigates the chemical activation of grinded CFA, as it improves the applicability of CFA in the aforementioned fields by considering alkali dissolution and fusion with subsequent functionalization.

2 Characterization of CFA

According to the European Standard EN450-1, CFA is categorized as "type II additions," which refers to finely divided materials with pozzolanic properties. These materials can be added to concrete to improve setting time, achieve early strength, or obtain specific desired properties (The European Committee for Standardization 2012). CFA is also defined as spherical particles primarily consisting of fine powder generated from the combustion of pulverized coal in thermal power plants. The ASTM Standard C-618 classifies CFA based on its chemical composition; if the weight percent of Al_2O_3, SiO_2, and Fe_2O_3 exceeds 70%, the CFA is classified as Class F CFA. On the other hand, if the weight percent of the same radicals falls in the range of 50 to 70 percent, the CFA is classified as cementitious Class C CFA (ASTM C 618-00 2000). CaO content also determines the class of CFA, as Class F CFA is characterized by a CaO content of less than 10% (by weight)

(Sear 2001). Anthracite or bituminous coal is the primary source of pozzolanic Class F CFA, whereas Class C CFA is produced by the combustion of sub-bituminous or lignite coal. Class C CFA possesses both cementitious and pozzolanic characteristics. Coal's primary minerals are alumino-silicates (including clay), carbonates, sulfides, and silica (quartz), which are also responsible for the composition of CFA.

CFA stands out as one of the most intricate materials to characterize due to its complex nature. Numerous CFA samples have been analyzed, revealing the presence of approximately 188 mineral groups across various samples (Vassilev et al. 2003; Vassilev and Vassileva 2005). Around 10 to 85% of coal residue contains CFA spherical particles (Stockel and Bridgewater 1984). CFA is an anthropogenic material that contain both amorphous and crystalline phases and a small amount of unburned carbon (Chancey et al. 2010).

According to Felekoglu et al. (2009), the fineness and specific surface area of CFA are the two most influential factors in its activity. The size of CFA particles is primarily determined by the primary grinding of the source coal, the combustion efficiency of the furnace in a thermal power plant, and even variation in electricity production (Lee et al. 2010). The average particle size of CFA has been determined by multiple researchers ≥300 μm (Rajak et al. 2017; Bhatt et al. 2019). The morphology of CFA is determined by utilizing scanning electron microscopy (SEM) analysis. The morphology of CFA depends on the temperature of coal combustion and the rate at which the combustion products cool down. Alumino-silicate ($Al_2SiO_5(OH)_4$), spheres (Cenosphere), iron-rich spheres (Ferrosphere), and calcium make up the majority of CFA.

Iron oxide (FeO) and amorphous $Al_2Si_2O_5(OH)_4$ make up the majority of the ferrosphere, while oxygen, sulfur, and phosphorus are associated with calcium (Kutchko and Kim 2006). Figure 1 depicts the typical CFA SEM monograph. The observed CFA was gathered at the Bokaro Thermal Power Station in Bokaro, Jharkhand, India. The energy-dispersive X-ray spectroscopy (EDX) and X-ray fluorescence (XRF) analyses in Table 1 represent the collected CFA sample from which unburned carbon has been removed. The oxide forms of Al, Ca, Fe, K, Mg, Si, and Ti are confirmed by EDS analysis, which is supported by XRF testing. The weight% of SiO_2 + Al_2O_3 + Fe_2O_3 is approximately 95.11, which is greater than 70% (from Table 1), confirming that the CFA belongs to Class F.

The CFA also contains a minor amount of unburned carbon. The presence of unburned carbon in CFA can be attributed primarily to inefficient combustion processes and the inherent properties of the coal itself (Bartoňová 2015). During combustion, coal char oxidation is slower than devolatilization (Al-Qayim et al. 2017). Thus, char reactivity is crucial in determining the amount of unburned carbon present in CFA. Other important factors include (a) the inorganic layer that forms on the surface of char during combustion, which prevents the diffusion of oxygen through the carbonaceous matrix (Wigley et al. 1997) and (b) the high specific heat capacity of mineral matter present in coal relative to organic grains, which causes it to heat up and combust more slowly than organic grains (Van Dyk et al. 2009).

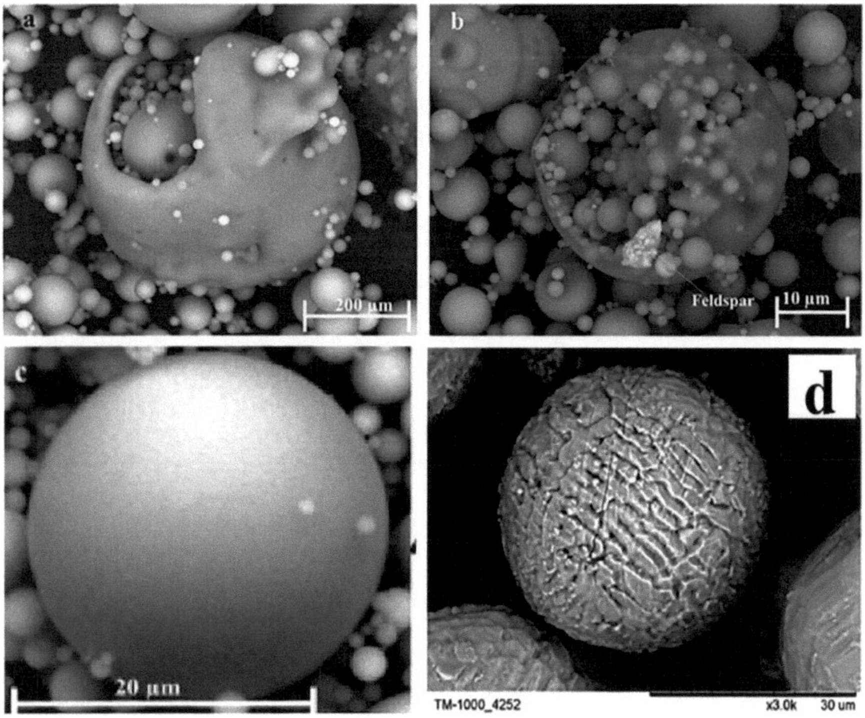

Fig. 1 SEM images of (**a**) CFA, (**b**) plerosphere rich in Al and Si, (**c**) cenosphere rich in alumino-silicate, (**d**) ferrosphere rich in ferrous particle (Figures are taken from Yadav and Fulekar 2020 which is adapted from Goodarzi and Sanei 2009; Sharonova et al. 2015)

Table 1 EDS and XRF analysis of CFA

Composition (wt%)								
Method	Sample	Mg	Si	Fe	Al	K	Ca	Ti
EDS	CFA	0.0	26.25 ± 1.8	5.41 ± 0.5	15.83 ± 1.5	0.69 ± 0.05	0.45 ± 0.01	1.88 ± 0.08
XRF	CFA	**MgO**	**SiO_2**	**Fe_2O_3**	**Al_2O_3**	**K_2O**	**CaO**	**TiO_2**
		0.0	56.22 ± 2.8	9.06 ± 0.58	29.83 ± 2.1	1.72 ± 0.15	0.54 ± 0.02	2.62 ± 0.1

In order to determine the particle size distribution and mean particle size of the newly acquired CFA from Bokaro Thermal Power Station (BTPS), dynamic light scattering (DLS) experiments were employed. The calculated mean particle size with 32% passing was found to be 134.5 μm (Fig. 2). This indicates that 68% of total fly ash was retained on a sieve with a 134.5 μm average opening.

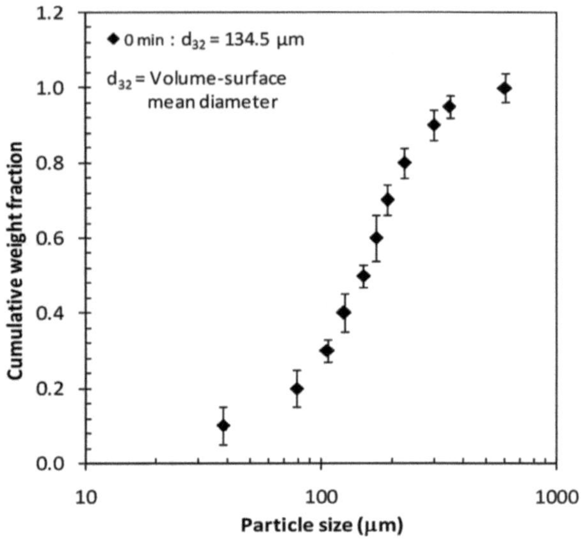

Fig. 2 DLS analysis of fresh CFA

Table 2 Particle size, specific gravity, and specific surface area of fresh fly ash

Physical properties of fresh fly ash	
Particle size	134.5 ± 8.54 μm
Specific gravity	2.25 ± 0.2
Specific surface area	7.31 ± 0.45 $m^2 g^{-1}$

The BET technique was utilized to determine the specific surface area of CFA, yielding a value of 7.31 ± 0.45 m^2/g. Additionally, the specific gravity of fly ash was measured to be 2.25 ± 0.2. Table 2 provides a list of previously measured values for particle size, specific gravity, and specific surface area.

3 Mechanical Activation

CFA can react with H_2O and CaO to produce a substance that behaves like cement; therefore, CFA is pozzolanic. Due to its pozzolanic nature, CFA has a wide range of applications in composite cement. It is also a basal material for the production of geopolymer and zeolite (Srivastava et al. 2023). SiO_2, Al_2O_3, Fe_2O_3, CaO, and varying quantities of carbon make CFA a strong alkaline substance with a pH between 10 and 13. When CFA and H_2O are combined, the surface becomes negatively charged, which can remove the toxic metal ion from the aqueous solution effectively (Xiyili et al. 2017). CFA is also a potential source of Al, Si, Fe, rare metals Ga, Ge, Se and rare earth metals La, Ce, Pr, and Nd, which are extracted via dissolution, solvent extraction, and/or precipitation (Srivastava et al. 2023). Due to the stable structure of crystalline phases (such as mullite and quartz) in CFA and the limited availability of the least stable and reactive amorphous phase, i.e., the glassy phase, the direct use of CFA in the aforementioned applications is restricted. Mechanical activation can change the crystalline phases of CFA into amorphous phases without altering the chemical compositions (Rajak et al. 2017; Grabias-Blicharz and Franus 2022).

Several researchers attempted to create ultrafine CFA particles by utilizing different mills and milling parameters (Bouzoubaa et al. 1997; Kumar et al. 2007; Paul et al. 2007; Aydin et al. 2010; Li et al. 2010; Kumar and Kumar 2011). CFA can be ultrafinely milled using a high-energy vibratory mill with balls and rings, a planetary ball mill, or an attrition mill, which improves CFA's reactivity by augmenting its specific surface area and amorphous content. Using a ball and vibratory mill, nano-sized CFA particles were effectively produced from coarse raw CFA; however, the grinding time was lengthy (Paul et al. 2007; Kumar and Kumar 2011). During milling, the crystalline nature is diminished, which enhances the bulk and surface activity of CFA containing a silanol (Si-OH) functional group (Paul et al. 2007). Various authors have conducted numerous studies on the mechanical activation of CFA through ultrafine milling. However, it is worth noting that the total duration of time necessary to achieve the desired particle size of CFA was relatively long, spanning from 2 to 96 h (Bouzoubaa et al. 1997; Paul et al. 2007; Aydin et al. 2010; Li et al. 2010). The crystalline nature of CFA (due to the presence of SiO_2, Al_2O_3, Fe_2O_3, Fe_2O_3, etc.) and various critical milling parameters such as material-to-ball ratio, jar filling volume, moisture content of CFA, and mishandling of the mill are the major factors for the increased grinding time (Mio et al. 2004; Sharma 2012; Patil and Anandhan 2015; Rajak et al. 2017).

In a study by Mio et al. (2004), the critical speed of jars and sun was calculated using a planetary ball mill with gibbsite powder as the milling material. Another study conducted by Patil and Anandhan (2015) investigated the wet grinding of CFA using a high-energy planetary ball mill. They examined the impact of the weight ratio of milling balls to CFA powder on the crystallite and particle size, as well as the specific surface area of the milled CFA. Moreover, Chen et al. (2015) investigated the effect of media density and rotation-to-revolution speed of jars on the morphology and particle size distribution of $CaCO_3$ powder using a planetary ball mill. Several attempts have been made to determine the energy consumption and milling time required to obtain the desired particle size using a ball mill by employing mathematical simulation and modeling with milling kinetic models based on population balance equations (Kapur and Fuerstenau 1987; Kapur et al. 2003; Bilgili and Scarlett 2005; Capece et al. 2011). Modeling and simulation of milling are not only useful for scale-up studies but also for selecting the optimal values of operating variables through process optimization. In addition, milling simulation enhances large-scale milling efficiency, allowing for the production of the desired end product at the lowest possible cost. Bilgili et al. (2006) used a laboratory-scale agitated media mill to determine the mechanisms and breakage kinetics for nano-sized pigment agglomerate using size-discrete population balance models. Several researchers have investigated the optimum operating parameters of planetary ball milling, such as power consumption, the angular speed of the jar and sun, and powder-to-ball loading, which aid in determining the physicochemical changes that occur in CFA during the grinding process (Watanabe 1999; Mio et al. 2004; Paul et al. 2007). Rajak et al. (2017) also attempted to calculate breakage parameters such as the breakage distribution parameter (Bi,j) and the specific rate of breakage (Si) using the direct- and back-calculation methods. CFA (Class F grade) was used to determine breakage kinetics, and the milling equipment was a high-energy planetary ball mill.

The CFA feed size was 120 μm which was milled to a mean particle size of 13.4 μm with a minimum particle size of 2.0 μm during the 2.5 h milling period. The reactivity of CFA was increased after milling as a result of the rise in specific surface area (i.e., from 7.31 m^2/g to 28.08 m^2/g) during milling, whereas crystallite size of SiO_2 was decreased by 40.1% and percent crystallinity of CFA was decreased by 26.12%. $B_{i,j}$ and S_i of CFA calculated using a direct experimental method were

$$S_i\left(\min^{-1}\right) = 0.00121 x_i^{1.08058}, B_{i,j} = 0.19\left(\frac{x_i}{x_{j+1}}\right)^{0.2656} + 0.81\left(\frac{x_i}{x_{j+1}}\right)^{2.4678}; n \geq i > j \geq 1 \text{ and } B_{i,j} = 1; i = j \text{ and } i = j + 1.$$

In order to simplify the tedious experimental procedures of the direct calculation method for determining the kinetic breakage parameters, the straightforward back-calculation method was employed for determining the values of $B_{i,j}$ and S_i. The values of S_i and $B_{i,j}$ calculated using the direct- and back-calculation method were quite similar. Finally, with the help of calculated $B_{i,j}$ and S_i values, the specific-energy breakage rate with particle size was also calculated. This laboratory milling simulation can be applicable to scale-up studies. While the primary focus was to examine the impact of mechanical activation on the surface and bulk properties of CFA so that CFA can be utilized efficiently in various industrial and commercial applications.

3.1 Microstructure Changes in Mechanically Activated CFA

CFA is composed of an outer reactive amorphous region and a less reactive crystalline interior. The CFA has been milled into smaller, more reactive particles as the glass content has increased. Figure 3 depicts the typical SEM images of raw CFA (RFA) and milled CFA (AMFA, VMFA, and CFA) produced by attrition mills (AMFA), vibrating mills (VMFA), and jet mills (CFA). The SEM image verifies the disintegration of large CFA particles and the formation of new, smaller particles. Figure 3 also depicts the corresponding Fourier-transformed infrared spectrometer (FTIR) analysis. Si-O-Si symmetric stretching (798 cm^{-1}), Si-O-Si bending (460 cm^{-1}), and Si/Al-O-Si asymmetric stretching (798, 913, 1090, and 1160 cm^{-1}) indicate the structural reconfiguration of CFA during milling (Lee and van Deventer 2002).

Figure 4 shows that milling enhances the amorphous content of CFA, which is supported by the X-ray diffraction plot of CFA. Figure 4 demonstrates that as

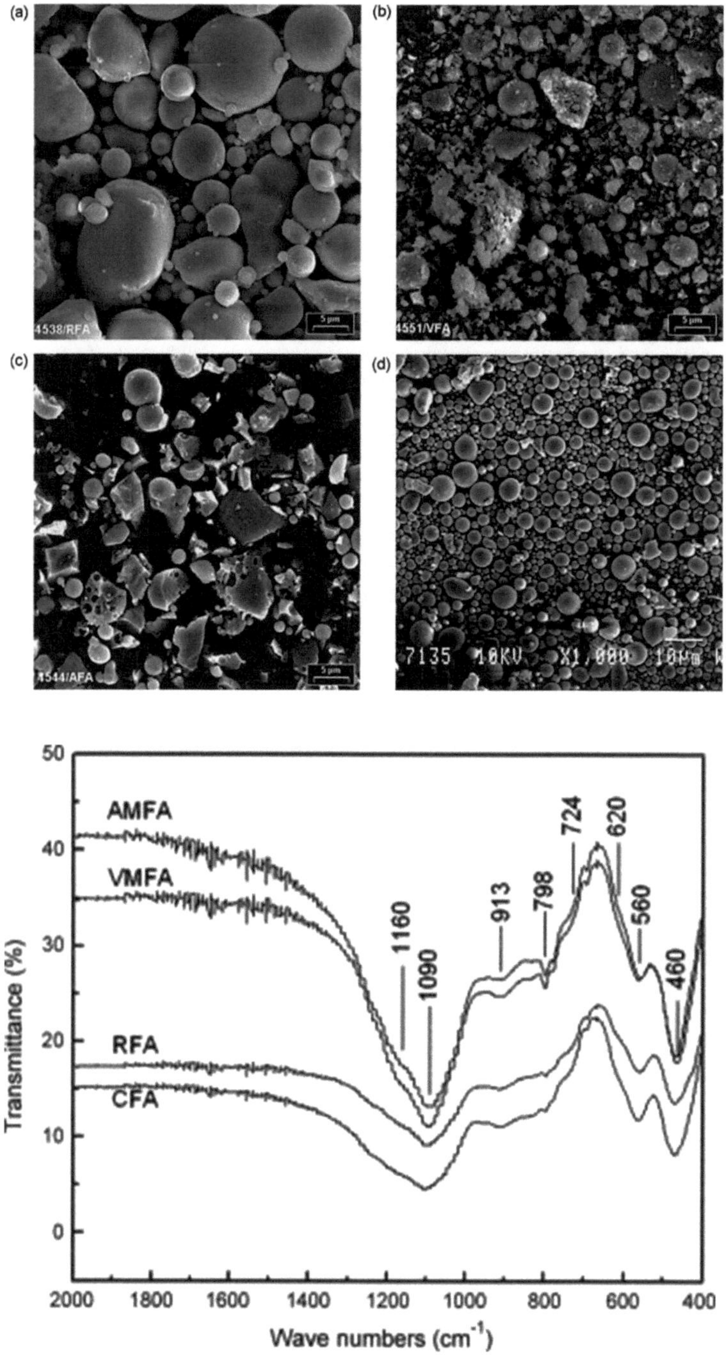

Fig. 3 SEM micrographs of (**a**) raw CFA (RFA), (**b**) vibratory milled CFA (VMFA), (**c**) attrition milled CFA (AMFA), (**d**) jet milled classified CFA (CFA) and corresponding Fourier transformed infrared spectrometer (FTIR) plot (adapted from Kumar et al. 2007, copyright Elsevier)

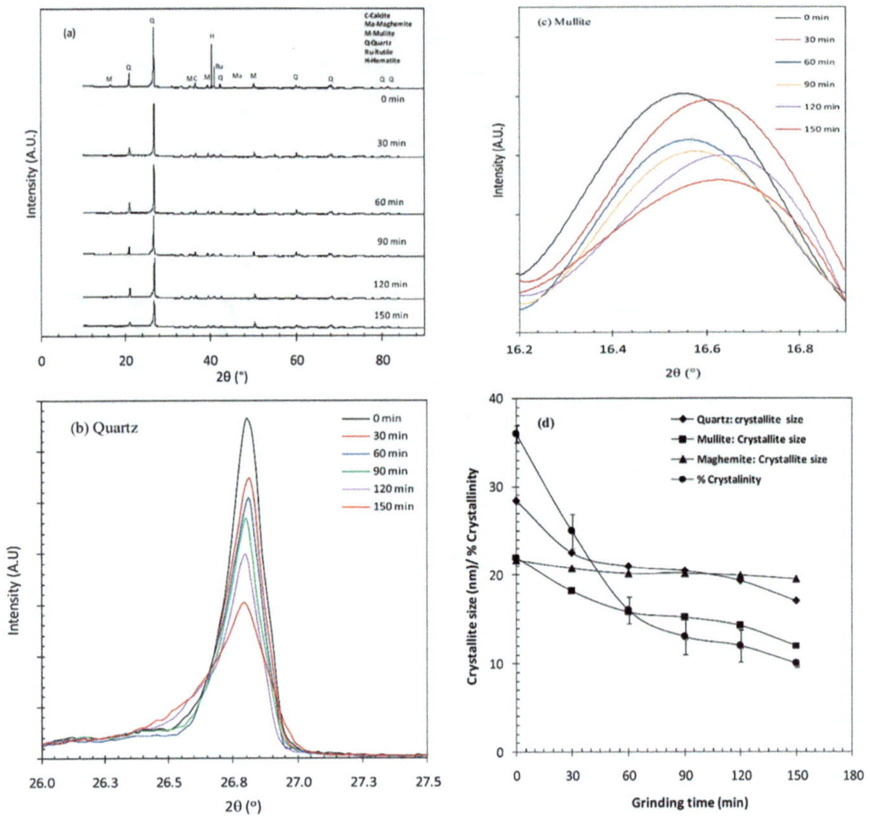

Fig. 4 (**a**) XRD of CFA grinded over 30 (1/2 h), 60(1 h), 90 (3/2 h), 120 (2 h), and 150 (5/2 h), (**b**) variation in the SiO_2 peak at $2\theta = 26.8°$ with grinding times (0, 30 (1/2 h), 60 (1 h), 90 (3/2 h), 120 (2 h), and 150 (5/2 h) minutes), (**c**) variation in the Al_2O_3 peak at $2\theta = 16.6°$ with milling times (0, 30, 60, 90, 120, and 150 min), and (**d**) variation in % crystallinity and crystallite size with respect to the milling time (adapted from Rajak et al. 2017 copyright Elsevier)

grinding progresses, the crystalline peaks diminish, indicating a notable rise in amorphous content and ultimately an enhanced reactivity. Figure 4 also illustrates how the crystalline apex of SiO_2 and Al_2O_3 decreases as grinding time progresses.

4 Chemical Activation of CFA

The mechanically activated CFA can be used more efficiently by treating the powdered CFA with compounds that provide a more active CFA surface. The surface-modified CFA can be readily functionalized, increasing its applications, such as heavy metal adsorption from wastewater and drilling fluid additives in the petroleum industry (Ahmaruzzaman 2010; Dash et al. 2022). When CFA is treated with alkali, amorphous constituents undergo dissolution, while crystalline phases remain unaffected (Brouwers and van Eijk 2002; Rowles and O'Connor 2009; Rajak et al. 2016). Several researchers investigated the aqueous solution alkali leaching mechanism of CFA spheres under the assumption that the amorphous portion of CFA was dissolved in an alkaline solution. The alkaline dissolution generates a porous "shell" and an unreacted core that are completely separated by a pointed boundary (Brouwers and van Eijk 2002; Rajak et al. 2016). In the alkaline dissolution of CFA, the diffusion of reacted SiO_2 and Al_2O_3 is fast relative to boundary movement, whereas the diffusivity of the dissolved amorphous phase is relatively sluggish. This form of process is controlled by diffusion and is known mathematically as a "pseudo-steady" or "quasi-stationary" process. The alkali dissolution of CFA can be described by Fick's law with invariable diffusivity is taken into account. Brouwers and van Eijk (2002) believed that the amorphous phase leaching in CFA using an aqueous solution of alkali falls under pseudo-steady-state conditions and that the effluent solution concentration remains unchanged during the course of the treatment, i.e., under conditions of infinite volume solution. In actual experimental conditions, however, it is difficult to maintain infinite-volume solution conditions, as the concentration of alkali gradually decreases during alkali leaching of CFA, i.e., finite solution volume conditions are attained. In addition, the reaction kinetics of dissolution for CFA leaching with an alkaline solution is more realistic under conditions of unsteadiness. Therefore, the batch scale investigation of alkali leaching for CFA can be viewed as a finite solution volume shell core model with a non-steady-state condition. Rajak et al. (2016) investigated the pseudo-steady and unsteady state shrinking core model for NaOH dissolution of CFA with shell–core behavior by assuming that the concentration of NaOH in the aqueous solution was significantly decreasing; this provides a better understanding of CFA leaching kinetics in the presence of NaOH containing aqueous solution. Using both "unsteady state" and "pseudo-steady state" models, the diffusivity of dissolved SiO_2 in the porous media ranged from 2.28×10^{11} to 2.96×10^{10} cm^2/s. This indicates that the alkali dissolution of CFA is a gradual process. Gautam et al. (2018) conducted an effective laboratory experiment by activating CFA with NaOH fusion and then substituting amine and epoxy in the activated CFA. The functionalized CFA aqueous solution was then measured for yield point gel strength, apparent and plastic viscosity, and filtration loss. The outcomes are comparable with the API-grade aqueous bentonite solution.

The primary objective of this endeavor is to utilize functionalized CFA as an oil well drilling fluid additive. Activating CFA with Na_2SiO_3, which forms a new geo-adsorbent via geopolymerization, increases the adsorption capacity of CFA as well. Geopolymerized CFA is effective at removing phenol, dyes, and heavy metal contaminants from effluent water (Ahmaruzzaman 2010; El Alouani et al. 2019). Geopolymers are gaining popularity in the construction industry as a result of their superior mechanical properties, thermal stability (remain unchanged up to 1200 °C), and lower Ca content, which provides superior resistance to acid attack and outperforms Portland cement (Palomo and Jimenez 2011; Schmücker and MacKenzie

Table 3 Geopolymer from CFA

Raw material	Operating conditions	Activator	Products and applications	References
CFA	Poly (sialate-siloxo) (PSS, SiO_2/Al_2O_3: 4) Temp.: 20–85 °C Time: 16–672 h	KOH and NaOH	The compressive strength of K-based geopolymers is 18.6 MPa, whereas Na-based geopolymer has 9.6 MPa. Suitable for making building blocks	Andini et al. (2008)
CFA with blast furnace slag	SiO_2/K_2O: 1.25 Temp.: 25–30 °C Time: 28 days	K_2SiO_3 and KOH	Synthesized geopolymer immobilize trace pollutants such as Th, U, Zr, Be, Co, Cu, Nb, Cr, Zr, Y, Ni, Sn, Cd, and Bi and rare earth elements during water leaching, which is a major concern in raw CFA and hazard to environment	Izquierdo et al. (2009)
CFA and blast furnace slag	SiO_2/K_2O: 1.25 Temp.: 25–30 °C and time: 28 days	K_2SiO_3	Synthesis of geopolymer with compressive strength 100 MPa, which is stronger and denser than Portland cement binders	Nugteren et al. (2009)
CFA	Al/Si: 0.43; Al/Na: 1,14; temp: 70 °C Time: 4 day	NaOH	Geopolymer has compressive strength 60 MPa. It is used as a fire-resistant coating on steel and binding material in construction materials to replace the Portland cement	Temuujin et al. (2010a, b)
CFA	CFA/NaOH: 1.25; temp: 25–105 °C Time: 4 days	NaOH	Ultrasonic alkaline dissolution. Potential adsorbent Pb from aqueous solution	Al-Harahsheh et al. (2015)
Class F CFA and copper mines tailings	Si/Al: 2; temp.: 60 °C, time: 24 h	NaOH	Alkaline dissolution. Use as a construction material	Zhang et al. (2011)
CFA	K_2SiO_3/KOH: 3.5; temp: 80 °C; time: 72 h	K_2SiO_3 and KOH	Potassium activation. Use as an adsorbent	Kaewmee et al. (2020)

2005). Numerous scientists have attempted to synthesize geopolymer from CFA using the activators outlined in Table 3. Table 3 verifies that alkalis are the primary activator of CFA geopolymerization. Silica and alumina dissolution occur during geopolymerization (Gollakota et al. 2019). Additionally, CFA finds application in the preparation of zeolites and mesoporous materials. However, zeolite synthesis is linked to potentially hazardous compounds derived from CFA. In hydrothermal and pyrothermal processes for zeolite preparation using CFA as a base material, Si and Al are extracted; consequently, there is a risk of contamination throughout the entire process, and there is also a risk of undesirable elements being mobilized into the environment (Ferrarini et al. 2016). Zeolite synthesis requires a substantial quantity of water. Geng et al. (2021) demonstrated that mechanochemical activation of CFA enhances Hg adsorption capacity.

Raw CFA has the lowest Hg removal efficacy, while mechanical-chemically activated CFA mechanical-chemically brominated CFA-impregnated brominated CFA has the highest adsorption capacity. The chemical activation of CFA with mineral acids also increases the silica content of the activated CFA, which is then used to prepare the catalyst (Khatri and Rani 2008). Rezaei and colleagues investigated a well-defined sodium salt (NaCl, $NaNO_3$, Na_2CO_3, and Na_2SO_4) roasting to decompose the alumino-silicate of CFA.

Ge, V, and Li can be extracted from roasted CFA when subjected to citric acid monohydrates ($C_6H_8O_7{\cdot}H_2O$), DL-malic acid ($C_4H_6O_5$), acetic acid ($C_2H_4O_2$), and oxalic acid dehydrate ($C_2H_2O_4{\cdot}2H_2O$) leaching. Citric acid leaching gives excellent results (Rezaei et al. 2022). $CaCl_2$, Na_2SO_4, $Ca(OH)_2$, Na_2SiO_3, and K_2SiO_3 are used to activate the CFA to increase its pozzolanic reactivity (Ahmaruzzaman 2010; Srivastava et al. 2023). The chemical activators provide an efficient and cost-effective technique. The addition of Na_2SO_4 and $CaCl_2{\cdot}\ 2H_2O$ can enhance the early and later strength of cement that incorporates CFA. Na_2SO_4 is particularly effective in improving early-age strength, while $CaCl_2{\cdot}\ 2H_2O$ exhibits greater effectiveness in enhancing later-age strength. CFA waste, which comprises diverse metal oxides and possesses thermal stability, serves as a cost-effective resource for catalyst preparation (Wang 2008). The major constituent of CFA is alumina-silicates, which makes it catalyst support for various reactions. CFA-supported CaO catalyst has application in Knoevenagel condensation reaction and transesterification of soybean oil (Kelechi et al. 2022). Chemically activated CFA loaded with sulfated zirconia can be used in benzylation reactions (Khatri et al. 2010). Nitric acid-activated CFA is used for the synthesis of heterogeneous Fenton-like catalysts, which can remove around 98% p-nitrophenol from wastewater (Park and Bae 2018). Sulfuric acid leached CFA followed by ethanol and organic silane treatment is used for the polystyrene-coated CFA catalyst preparation, which can be used for the synthesis of alkyl levulinates (a good solvent and additives) (Tian et al. 2023).

5 Conclusions

Raw CFA is a finely grained and powdery material composed of particulates, which exhibit thermal stability. It is obtained by collecting it from flue gases through the utilization of an electrostatic precipitator. It is composed of metal and metalloid oxides. CFA also contains heavy metals such as Mn, Hg, Be, Zn, Sb, Cu, Co, Sr, F, and As, and long-term, large-scale storage may have severe environmental consequences. Through the leaching of these heavy metals, CFA has the potential to negatively impact the soil and adjacent groundwater, as well as the plants, food chain, and local environment. However, the predominant components found in CFA include Al_2O_3,SiO_2, and CaO, and its physical properties, such as amorphous spherical fine powder particles (hollow or solid), lead to various applications in cement additives, concrete making, geopolymer and geolite production, catalyst preparation, corrosion-resistant material, and wastewater treatment after mechanical activation and/or chemical activation. Size reduction (mechanical activation), which provides surface and bulk modification with the use of suitable compounds (chemical activation), can increase CFA utilization. Mechanical and chemical activation significantly modifies the morphology of CFA. Mechanical activation of CFA also increases the area to mass ratio and amorphous content, which renders the CFA more reactive and increases the efficiency of its use. The chemistry of CFA activation is limited to surface modification and bulk functionalization. CFA's surface modification renders it reactive and suitable for a variety of productive reactions. The surface is modified by utilizing various alkali and/or acid dissolutions or occasionally alkali fusion when less reaction time is required. The functionalization of chemically activated CFA generates polymer-like material that can be utilized in various catalyst preparation applications. As its aqueous solution possesses the same rheological and filtration loss properties as the aqueous solution of AP grade bentonite, the functionalization of modified CFA may be utilized as a drilling fluid additive. However, mechanical activation of CFA is an energy-intensive procedure, whereas chemical activation generates numerous hazardous by-products. Consequently, substantial research is still required to optimize these processes to be cost effective and environmentally beneficial.

Acknowledgments The work was supported by Indian Institute of Technology (Indian School of Mines), Dhanbad, India, and Institute of Engineering, Pulchowk Campus, Tribhuvan University, Nepal.

References

Ahmaruzzaman A (2010) A review on the utilization of fly ash. Prog Energ Combust Sci 36:327–363

Al-Harahsheh MS, AlZboon K, Al-Makhadmeh L, Hararah M, Mahasneh M (2015) Fly ash based geopolymer for heavy metal removal: a case study on copper removal. J Environ Chem Eng 3: 1669–1677

Al-Qayim K, Nimmo W, Hughes K, Pourkashanian M (2017) Kinetic parameters of the intrinsic reactivity of woody biomass and coal chars via thermogravimetric analysis. Fuel 210:811–825

Andini S, Cioffi R, Colangelo F, Grieco T, Montagnaro F, Santoro L (2008) Coal fly ash as raw material for the manufacture of geopolymer-based products. Waste Manag 28:416–423

ASTM C 618-00 (2000) Standard specification for coal fly ash and raw or calcined natural pozzolan for use as a mineral admixture in concrete. Annual Book of ASTM Standards, Philadelphia

Aydin S, Karatay C, Baradan B (2010) The effect of grinding process on mechanical properties and alkali-silica reaction of fly ash incorporated cement mortar. Powder Tecnol 197:68–72

Bartoňová L (2015) Unburned carbon from coal combustion ash: an overview. Fuel Process Technol 134:136–158

Bentz DP (2010) Powder additions to mitigate retardation in high-volume fly ash mixtures. ACI Mater J 107:508–514

Bentz DP, Ferraris CF (2010) Rheology and setting of high volume fly ash mixtures. Cem Concr Compos 32:265–270

Bhatt A, Priyadarshini S, Mohanakrishnan AA, Abri A, Sattler M, Techapaphawit S (2019) Physical, chemical, and geotechnical properties of coal fly ash: a global review. Case Stud Constr Mater 11:e00263

Bilgili E, Scarlett B (2005) Population balance modeling of non-linear effects in milling processes. Powder Technol 153:59–71

Bilgili E, Hamey R, Scarlett B (2006) Nano-milling of pigment agglomerates using a wet stirred media mill: elucidation of the kinetics and breakage mechanisms. Chem Eng Sci 61:149–157

Bilodeau A, Malhotra V (2000) High-volume fly ash system: concrete solution for sustainable development. ACI Mater J 97:41–48

Blicharz EG, Panek R, Franus M, Franus W (2022) Mechanochemically assisted coal fly ash conversion into zeolite. Materials 15:7174

Borm JAP (1997) Toxicity and occupational health hazards of coal fly ash (CFA). A review of data and comparision to coal mine dust. Ann Occup Hyg 41:659–676

Bouzoubaa N, Zhang MH, Bilodeau A, Malhotra VM (1997) The effect of grinding on the physical properties of fly ashes and a Portland cement clinker. Cem Concr Res 27:1861–1874

Brouwers HJH, Van Eijk RJ (2002) Reactivity of fly ash: extension and application of a shrinking core model. J Mater Sci 37:2129–2141

Capece M, Bilgili E, Dave R (2011) Identification of the breakage rate and distribution parameters in a non-linear population balance model for batch milling. Powder Technol 208:195–204

Carlson CL, Adriano DC (1993) Environmental impacts of coal combustion residues. J Environ Qual 22:27–247

Chancey RT, Stutzman P, Juenger MCG, Fowler DW (2010) Comprehensive phase characterization of crystalline and amorphous phases of a class F fly ash. Cem Concr Res 40:146–156

Chen Y, Lian X, Li Z, Zheng S, Wang Z (2015) Effects of rotation speed and media density on particle size distribution and structure of ground calcium carbonate in a planetary ball mill. Adv Powder Technol 26:505–510

Dash S, Panda L, Mohanty I, Gupta P (2022) Comparative feasibility analysis of fly ash bricks, clay bricks and fly ash incorporated clay bricks. Mag Civil Eng 115:11502

El Alouani M, Alehyen S, El Achouri M, Taibi MH (2019) Comparative study of the adsorption of micropollutant contained in aqueous phase using coal fly ash and activated coal fly ash: kinetic and isotherm studies. Chem Data Collec 23:100265

Felekoglu B, Trkel S, Kalyoncu H (2009) Optimization of fineness to maximize the strength activity of high-calcium ground fly ash—Portland cement composites. Construct Build Mater 23:2053–2061

Ferrarini SF, Cardoso AM, Paprocki A, Pires M (2016) Integrated synthesis of zeolites using coal fly ash: element distribution in the products, washing waters and effluent. J Braz Chem Soc 27: 2034–2045

Gautam S, Guria C, Rajak DK, Pathak AK (2018) Functionalization of fly ash for the substitution of bentonite in drilling fluid. J Petrol Sci Eng 166:63–72

Geng X, Duan Y, Zhao S, Hu J, Zhao W (2021) Mechanism study of mechanochemical bromination on fly ash mercury removal adsorbent. Chemosphere 274:129637

Gollakota AR, Volli V, Shu CM (2019) Progressive utilisation prospects of coal fly ash: a review. Sci Total Environ 672:951–989

Goodarzi F, Sanei H (2009) Plerosphere and its role in reduction of emitted fine fly ash particles from pulverized coal-fired power plants. Fuel 88:382–386

Grabias-Blicharz E, Franus W (2022) A critical review on mechanochemical processing of fly ash and fly ash-derived materials. Sci Total Environ 860:160529

Izquierdo M, Querol X, Davidovits J, Antenucci D, Nugteren H, FernándezPereira C (2009) Coal fly ash-slag-based geopolymers: microstructure and metal leaching. J Hazard Mater 166:561–566

Juenger MCG, Winnefeld F, Provis JL, Ideker JH (2011) Advances in alternative cementitious binders. Cem Concr Res 41:1232–1243

Kaewmee P, Song M, Iwanami M, Tsutsumi H, Takahashi F (2020) Porous and reusable potassium-activated geopolymer adsorbent with high compressive strength fabricated from coal fly ash wastes. J Clean Prod 272:122617

Kapur PC, Fuerstenau DW (1987) Energy-size reduction laws revisited. Int J Miner Process 20:45–57

Kapur PC, Fuerstenau DW, De A (2003) Modelling breakage kinetics in various dry comminution systems. KONA Powder Part J 21:121–132

Kelechi SE, Adamu M, Uche OAU, Okokpujie IP, Ibrahim YE, Obianyo II (2022) A comprehensive review on coal fly ash and its application in the construction industry. Cogent Eng 9: 2114201

Khatri C, Rani A (2008) Synthesis of a nano-crystalline solid acid catalyst from fly ash and its catalytic performance. Fuel 87:2886–2892

Khatri C, Jain D, Rani A (2010) Fly ash-supported cerium triflate as an active recyclable solid acid catalyst for Friedel–Crafts acylation reaction. Fuel 89:3853–3859

Kumar S, Kumar R (2011) Mechanical activation of fly ash: effect on reaction structure and properties of resulting geopolymer. Ceram Int 37:533–541

Kumar R, Kumar S, Mehrotra SP (2007) Towards sustainable solutions for fly ash through mechanical activation. Resour Conserv Recy 52:157–179

Kutchko BG, Kim AG (2006) Fly ash characterization by SEM-EDS. Fuel 85:2537–2544

Lee WKW, Van Deventer JSJ (2002) Structural reorganisation of class F fly ash in alkaline silicate solutions. Colloids Surf A Physicochem Eng Aspects 211:49–66

Lee HK, Kim HK, Hwang EA (2010) Utilization of power plant bottom ash as aggregates in fiber-reinforced cellular concrete. Waste Manag 30:274–284

Li MG, Sun CJ, Gau SH, Chuang CJ (2010) Effects of wet ball milling on lead stabilization and particle size variation in municipal solid waste incinerator fly ash. J Hazard Mater 174:586–591

Mio H, Kano J, Saito F, Kaneko K (2004) Optimum revolution and rotational directions and their speeds in planetary ball milling. Int J Miner Process 74:85–92

Nidheesh PV, Kumar MS (2019) An overview of environmental sustainability in cement and steel production. J Clean Prod 231:856–871

Nugteren HW, Butselaar-Orthlieb VCL, Izquierdo M (2009) High strength geopolymers produced from coal combustion fly ash. Global NEST J 11:155–161

Page AL, Elseewi A, Straughan IR (1979) Physical and chemical properties of fly ash from coal-fired power plants with reference to environmental impacts. In: Gunther FA, Gunther JD (eds) Residue reviews: residues of pesticides and other contaminants in the total environment. Springer, New York

Palomo, A, Jimenez, FA (2011) Alkaline activation, procedure for transforming fly ash into new materials. Part I: Application. World of coal ash (WOCA) conference, May 9–12, 2011 in Denver, CO, USA

Park J, Bae S (2018) Formation of Fe nanoparticles on water-washed coal fly ash for enhanced reduction of p-nitrophenol. Chemosphere 202:733–741

Patil AG, Anandhan S (2015) Influence of planetary ball milling parameters on the mechano-chemical activation of fly ash. Powder Technol 281:151–158

Paul KT, Satpathy SK, Manna I, Chakraborty KK, Nando GB (2007) Preparation and characterization of nano structured materials from fly ash: a waste from thermal power stations, by high energy ball milling. Nanoscale Res Lett 2:397–404

Pietersen HS (1990) Reactivity of fly ash at high pH. Mater Res Soc Symp Proc 178:139–157

Provis JL, Van Deventer JS (2013) Alkali activated materials: state-of-the-art report, RILEM TC 224-AAM, vol 13. Springer Science & Business Media, South Yorkshire

Rajak DK, Guria C, Ghosh R, Agarwal S, Pathak AK (2016) Alkali assisted dissolution of fly ash: a shrinking core model under finite solution volume condition. Int J Miner Process 155:106–117

Rajak DK, Raj A, Guria C, Pathak AK (2017) Grinding of class-F fly ash using planetary ball mill: a simulation study to determine the breakage kinetics by direct-and back-calculation method. S Afr J Chem Eng 24:135–147
Rezaei H, Shafaei SZ, Abdollahi H, Shahidi A, Ghassa S (2022) A sustainable method for germanium, vanadium and lithium extraction from coal fly ash: sodium salts roasting and organic acids leaching. Fuel 312:122844
Rowles MR, O'Connor BH (2009) Chemical and structural microanalysis of aluminosilicate geopolymers synthesized by sodium silicate activation of metakaolinite. J Am Ceram Soc 92 (10):2354–2361. https://doi.org/10.1111/j.1551-2916.2009.03191.x
Schmücker M, MacKenzie KJ (2005) Microstructure of sodium polysialatesiloxo geopolymer. Ceram Int 31:433–437
Sear LKA (2001) Properties and use of coal fly ash: a valuable industrial by-product. Thomas Telford, London
Sharma A (2012) Modification in properties of fly ash through mechanical and chemical activation. Am Chem Sci J 2:177–187
Sharonova OM, Anshits NN, Fedorchak MA, Zhizhaev AM, Anshits AG (2015) Characterization of ferrospheres recovered from high-calcium fly ash. Energy Fuel 29:5404–5414
Singh RK, Gupta NC, Guha BK (2016) Fly ash disposal in ash ponds: a threat to ground water contamination. J Inst Eng India Ser A 97:255–260
Srivastava RR, Rajak DK, Ilyas S, Kim H, Pathak P (2023) Challenges, regulations, and case studies on sustainable management of industrial waste. Fortschr Mineral 13:51
Stockel RF, Bridgewater NJ (1984) Coal ash fertilizer compositions. United States Patent, 4469503
Temuujin J, Minjigmaa A, Rickard W, Lee M, Williams I, vanRiessen A (2010a) Fly ash based geopolymer thin coatings on metal substrates and its thermal evaluation. J Hazard Mater 180: 748–752
Temuujin J, VanRiessen A, MacKenzie KJD (2010b) Preparation and characterisation of fly ash based geopolymer mortars. Construct Build Mater 24:1906–1910
The European Committee for Standardization (2012) Fly ash for concrete—part 1: definition, specification and conformity criteria. The European Committee for Standardization (CEN), EN 450-1. Brussels, Belgium
Tian Y, Zhu X, Zhou S, Zhao W, Xu Q, Liu X (2023) Efficient synthesis of alkyl levulinates fuel additives using sulfonic acid functionalized polystyrene coated coal fly ash catalyst. J Bioresour Bioprod 8:198–213
Van Dyk JC, Benson SA, Laumb ML, Waanders B (2009) Coal and coal ash characteristics to understand mineral transformations and slag formation. Fuel 88:1057–1063
Vassilev SV, Vassileva CG (2005) Methods for characterization of cmposition of fly ashes from coal-fired power stations: a critical overview. Energy Fuel 19:1084–1098
Vassilev SV, Menendez R, Alvarez D, Diaz-Somoano M, Martinez-Tarazona MR (2003) Phase-mineral and chemical composition of coal fly ashes as a basis for their multicomponent utilization. 1. Characterization of feed coals and fly ashes. Fuel 82:1793–1811
Wang S (2008) Application of solid ash based catalysts in heterogeneous catalysis. Environ Sci Technol 42:7055–7063
Watanabe H (1999) Critical rotation speed for ball-milling. Powder Technol 104:95–99
Wigley F, Williamson J, Gibb WH (1997) The distribution of mineral matter in pulverised coal particles in relation to burnout behaviour. Fuel 76:1283–1288
Xiyili H, Çetintaş S, Bingöl D (2017) Removal of some heavy metals onto mechanically activated fly ash: modeling approach for optimization, isotherms, kinetics and thermodynamics. Process Saf Environ Prot 109:288–300
Yadav VK, Fulekar MH (2020) Advances in methods for recovery of ferrous, alumina, and silica nanoparticles from fly ash waste. Ceramics 3:384–420
Zhang L, Ahmari S, Zhang J (2011) Synthesis and characterization of fly ash modified mine tailings-based geopolymers. Construct Build Mater 25:3773–3781

Environmental Damages Due to Mismanagement of Municipal Solid Waste

Dalia Carbonel, Yordin Garriazo, Mary Mayhua, Sara Orozco, and M. S. S. R. Tejaswini

Abstract Solid waste's generation rate depends on any country's urbanization and economic development. Among all types of SW processes, management and handling of municipal solid waste (MSW) is one of the significant concerns. Generally, in developing countries, the discarded MSW is either being dumped in landfills or directly combusted in open air. This unscientific manner of treating solid wastes causes environmental pollution and global climate change besides posing a threat to human health. Moreover, it is projected that there would be an increase in population of ~3.4 billion tons by 2050, thereby adding to the new challenges in managing solid waste due to limited landfill space and lack of technologies in treating them. Thus, accordingly, this chapter provides an overview of various sources, environmental damages, and the potential solutions that can help address these environmental damages caused by MSW mismanagement. Sustainable management mainly includes waste reduction, composting, recycling, and energy recovery. The effective utilization of these technologies directly or indirectly depends on composition, geographical location, attributes, funding accessibility, and externalities associated with each method.

Keywords Environmental damages · Mismanagement · Municipal solid waste · Waste management practices

D. Carbonel (✉)
Faculty of Environmental Engineering, National University of Engineering, Lima, Peru
e-mail: dcarbonelr@uni.pe

Y. Garriazo · M. Mayhua · S. Orozco
Lima Compost, Lima, Peru

M. S. S. R. Tejaswini
Department of Environmental Science and Engineering, SRM University, Amaravathi, Andhra Pradesh, India

P. Pathak et al. (eds.), *Anthropogenic Environmental Hazards*,
https://doi.org/10.1007/978-3-031-41013-0_8

1 Introduction

In developing countries, increased urbanization and industrialization are leading to the increased consumption and production of different types of solid wastes. MSW mainly comprises organic waste (food and green waste), paper waste (newspapers and cardboard), plastic waste (bottles, packaging materials, and disposable items), glass waste (broken glassware and colored glasses), metals (aluminum cans, steel containers, and foils), and the inorganic inert waste (nondegradable and nonrecyclable materials) as shown in Fig. 1a. Globally, it is currently estimated that two billion tons of MSW are produced, with nearly one-third (33%) of it not being collected by municipalities (Kaza et al. 2018). On average, according to the World Bank (Kaza et al. 2018), one person can generate from 0.11 to 4.54 kg of waste per day (Table 1). This figure tends to approximate the maximum in regions like Europe and Central Asia and North America, and it is lower in Sub-Saharan Africa and East Asia and Pacific.

According to the What a Waste 2.0 report (Kaza et al. 2018), the composition of MSW in different locations is shown in Table 2. As can be observed from the table, the MSW is primarily composed of organic wastes which is nearly around 50%. The organic and biodegradable portions of MSW are subjected to composting (5%), and less than 9% of MSW is recycled across various parts of the world. However, this

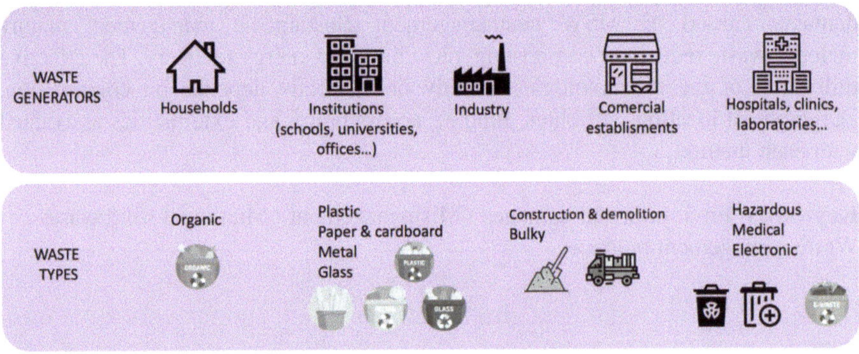

Fig. 1 MSW waste generators and types

Table 1 Ranges of average national waste generation (kg/capita/day) by region (Kaza et al. 2018)

Region	2016 Average	Min	25th Percentile	75th Percentile	Max
Sub-Saharan Africa	0.46	0.11	0.35	0.55	1.57
East Asia and Pacific	0.56	0.14	0.45	1.36	3.72
South Asia	0.52	0.17	0.32	0.54	1.44
Middle East and North Africa	0.81	0.44	0.66	1.40	1.83
Latin America and Caribbean	0.99	0.41	0.76	1.39	4.46
Europe and Central Asia	1.18	0.27	0.94	1.53	4.45
North America	2.21	1.94	2.09	3.39	4.54

Table 2 Ranges of average waste composition by region (Kaza et al. 2018)

Region	Food and green	Glass	Metal	Other	Paper and cardboard	Plastic	Rubber and leather	Wood
Sub-Saharan Africa	43%	3%	5%	30%	10%	8.6%	<1%	<1%
East Asia and Pacific	53%	2.6%	3%	12%	15%	12%	<1%	2%
South Asia	57%	4%	3%	15%	10%	8%	2%	1%
Middle East and North Africa	58%	3%	3%	8%	13%	12%	2%	1%
Latin America and Caribbean	52%	4%	3%	15%	13%	12%	<1%	<1%
Europe and Central Asia	36%	8%	3%	21%	18.6%	11.5%	<1%	1.6%
North America	28%	4.5%	9.3%	3.6%	28%	12%	9%	5.6%
Global	44%	5%	4%	14%	17%	12%	2%	2%

percentage is lower in more developed regions such as Europe and Central Asia and North America, where it is 36% and 28%, respectively. In North America, the lower percentage of green and food waste results in a higher proportion of metal, paper and cardboard, rubber and leather, and wood waste compared to the global average. Additionally, according to the World Bank's What a Waste 2.0 report, the per capita generation (kg/cap/day) of industrial, agricultural, construction, hazardous, medical, and electronic waste is 12.73, 3.35, 1.68, 0.32, 0.25, and 0.02, respectively (Kaza et al. 2018).

In the natural environment, direct combustion of MSW causes gaseous emissions that is mainly composed of hazardous pollutants including particulate matters, oxides of sulfur, carbon, and nitrogen, and volatile matters in the air. In the openly dumped areas, the decomposition of biodegradable organic waste produces huge amounts of methane (greenhouse) gas, which directly contributes to global warming. It is estimated that greenhouse gas emissions generated by the disposal of food waste amount to 887 Mt/year of CH_4, 77 Mt/year of N_2O, and 996 Mt/year of CO_2eq (Tubiello et al. 2021). Furthermore, presence of high organic contents in MSW attracts various microbial pathogens that cause several chronic diseases to rag pickers and nearby people living in that area. These emissions have significant environmental impacts, highlighting the urgent need for effective waste management strategies that prioritize the reduction, reuse, and recycling of MSW.

Apart from causing air pollution, improper disposal of MSW also pollutes water bodies and soil. The leachate through the solid waste seeps into groundwater, contaminating it with hazardous chemicals and pathogens. When waste is not disposed of properly, it can also end up in waterways, causing harm to aquatic life and making water unfit for human consumption. Soil pollution mainly occurs because of the leakage of hazardous chemicals and heavy metals into the soil, making it unfit for farming and other uses. In addition, illegal dumping of waste in open areas can cause soil erosion and other environmental problems.

The mismanagement of municipal solid waste can have significant and lasting impacts on the environment. It is essential that proper waste management practices be implemented to minimize these impacts and prevent further damage. By adopting best practices, such as waste reduction, recycling, and safe disposal, we can ensure that our cities are healthy and sustainable for generations to come. Proper waste management practices are essential to prevent environmental damage and ensure a healthy and sustainable future. The improper disposal of waste can have severe impacts on the environment, including air, water, and soil pollution, and can also contribute to climate change.

2 MSW Characteristics and Types

2.1 MSW Characteristics

MSW is a complex mixture of materials, including food waste, paper, plastics, metals, glass, and textiles, as well as hazardous waste such as batteries and electronic devices. It is highly heterogeneous in the nature which makes the sustainable management of MSW more challenging. The composition of MSW can vary depending on factors such as geography, climate, population density, and cultural practices. This makes it challenging to develop a one-size-fits-all approach to MSW management. Technical reasons for the challenges in MSW mismanagement can include issues with waste separation, transportation, and disposal methods. Waste separation can be challenging due to inadequate infrastructure, lack of education and awareness among citizens, and limited funding for waste management programs. Issues with transportation infrastructure, such as poor road conditions and limited access to collection sites, can lead to delays and increased costs (Kumar 2016; Nanda and Berruti 2021).

2.2 Types of MSW Generated in Urban Areas

MSW generated in urban areas can be classified into different types based on their composition and potential environmental impact. The main types of MSW generated in urban areas are organic, plastic, paper, and carboard, metal, glass, hazardous, electronic waste (e-waste), construction and demolition (C&D), and bulky and medical waste (Kumar 2016; Reddy 2011) (Fig. 1).

2.2.1 Organic Waste

Organic waste is a common type of MSW generated in urban areas, which refers to biodegradable waste derived from plants and animals. Organic wastes mainly include fruit and vegetable scraps, meat and dairy products, bakery waste, coffee grounds, grass clippings, leaves, and tree trimmings. The percentage of organic matter in these wastes can vary widely depending on the specific material, with food waste typically having a high organic content ranging from 70 to 90% organic matter and yard waste having estimates ranging from 50 to 80% organic matter (FAO 2014). Overall, organic waste is generally defined as any waste material that originates from a living organism and is biodegradable, including biodegradable materials from the food and agricultural industry, as well as other biodegradable waste such as yard waste and paper products.

2.2.2 Plastic Waste

Plastic waste is a major type of MSW generated in urban areas. Plastics can be classified into different categories based on their recyclability and degradability. Recyclable plastics are those that can be collected, sorted, and processed into new products. These mainly include polyethylene terephthalate (PET), high-density polyethylene (HDPE), polyvinyl chloride (PVC), and polypropylene (PP). Nonrecyclable plastics are those that cannot be effectively recycled due to technical or economic reasons. Examples of nonrecyclable plastics include polystyrene (PS), multilayer plastics, and low-density polyethylene (LDPE). On the other hand, degradable plastics are those that break down into smaller pieces over time but may not completely biodegrade. Examples of degradable plastics include oxo-degradable and photodegradable plastics. Nondegradable plastics are those that do not biodegrade or degrade significantly in the environment (Tejaswini et al. 2022). Examples of nondegradable plastics include most types of plastics, such as PET, HDPE, LDPE, and PP. It is important to note that some plastics may have properties that fall into multiple categories, and the proper classification of plastics can vary depending on the specific context and definitions used.

2.2.3 Paper and Carboard Waste

Paper and cardboard waste refers to waste products derived from paper and cardboard, including newspapers, magazines, cardboard boxes, and packaging materials. Paper and cardboard production is generally considered energy-intensive and the amount of energy consumed can vary depending on factors such as the production process and location. In the United States, the energy consumption for paper and cardboard production ranges from 11 to 22 GJ/MT (EPA 2022), while in the European Union, it is estimated to be around 12–15 GJ/MT (EC 2018). The energy consumption includes both direct and indirect energy use, and factors such as the efficiency of the production process and the use of renewable energy sources can also affect it.

2.2.4 Glass Waste

Glass waste denotes to waste products derived from glass, including beverage containers, food jars, and glass packaging materials. Glass waste is generated from households, businesses, and industries. The nonbiodegradable nature of glass waste and its potential to harm the environment make it a significant concern in managing municipal solid waste (MSW).

2.2.5 Hazardous and Electronic Waste

Hazardous and electronic wastes refer to waste products that have the potential to cause harm to human health or the environment and are derived from households, businesses, and industries. Improper management of these waste products poses significant concerns in the management of MSW. Common examples of hazardous and electronic waste include batteries, electronic devices, pesticides, chemicals, medical waste, and electronic devices such as computers, cell phones, and televisions. These waste products contain harmful substances such as lead, mercury, and arsenic, which can cause serious health problems if not managed properly.

2.2.6 C&D and Bulky Waste

C&D waste denotes to waste products derived from construction, renovation, and demolition activities such as bricks, concrete, wood, steel, and other building materials. Bulky waste indicates to large items such as furniture, appliances, mattresses, and other household items that are too big for regular waste disposal systems. C&D and bulky waste are generated from households, businesses, and industries.

3 Current Scenario of MSW Management

Analyzing the current state of solid waste management, like the study of any human activity, is a broad, polarized, and contradictory topic. It is broad in the sense that waste management is implicit in any human activity. It is polarized due to the inequalities inherent in community life, particularly in urban areas. In these areas, the central and more affluent zones tend to have some good waste management practices, while the peripheral and marginalized areas, due to a lack of resources among other reasons, tend to have poor solid waste management practices.

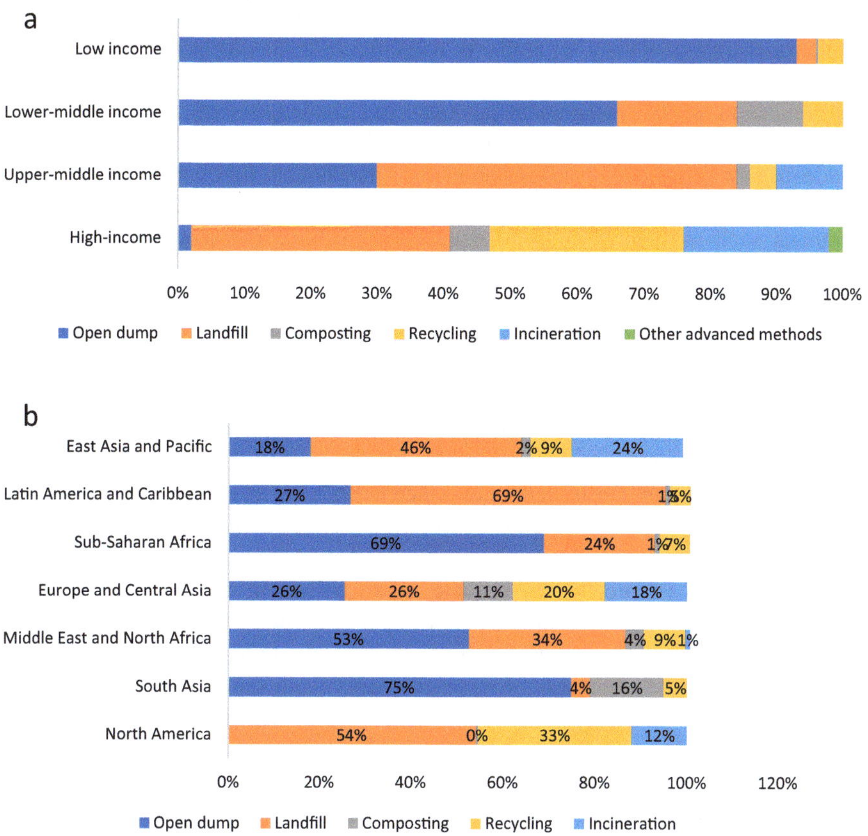

Fig. 2 Disposal methods by income level (**a**) and region (**b**) (Kaza et al. 2018)

Furthermore, it can also be contradictory when the adoption of modern practices is counterproductive to local practices. There is the example of the use of waste compactors in urban areas of Lima. The waste composition in this city, like many located in Latin American countries, is more than 50% organic. This means that when waste is compacted, leachate is generated and inadvertently discharged onto the road. Waste compactors are useful for predominantly dry and compactible waste, such as plastics, cardboard, or cans. This inequality and variability make it very interesting to review the extremes of the situation.

Figure 2 exemplifies these differences. High-income countries tend to have more friendly environment disposal methods such as recycling and incineration (which can cause zero pollution if proper managed). Low-income countries, however, usually open dumps their waste, leading to several environmental problems. A similar scenario occurs in the regions that were developed countries (i.e., North America) and have more sustainable waste management practices than developing countries (i.e., Latin America).

Table 3 provides a summary of the waste generation rate, collection coverage, main waste treatment disposal methods, recycling rates, and main MSWM problems in Bangladesh, China, Brazil, Switzerland, Israel, and the USA. It highlights the differences in waste management practices and challenges across these countries, with some countries facing more significant challenges such as limited landfill space, inadequate waste disposal infrastructure, and inadequate funding. Other countries have achieved high collection coverage rates and use more advanced waste treatment methods such as incineration and anaerobic digestion. The recycling rates in all countries could be improved, and there is a need for better education and enforcement of MSWM regulations and laws.

Bangladesh is a densely populated country with limited resources for waste management. As a result, many areas suffer from inadequate waste collection and disposal services, leading to uncontrolled dumping and pollution. China is the world's largest generator of municipal solid waste. However, the country has made significant efforts to improve its waste management system in recent years, including promoting waste reduction, increasing waste-to-energy capacity, and enforcing stricter regulations. Brazil has a relatively advanced waste management system compared to many other developing countries. However, the country still faces challenges in providing adequate waste collection and disposal services, particularly in rural and low-income areas. In recent years, there have been efforts to increase recycling and promote more sustainable waste management practices in Brazil, including the expansion of waste-to-energy facilities.

Analyzing the current state of solid waste management is a complex task that requires considering the broad, polarized, and contradictory aspects of waste management practices, particularly in urban areas. The case of Lima exemplifies how the adoption of modern practices can be counterproductive when they do not take into account the local context and the composition of the waste. Therefore, it is necessary to develop strategies that are adapted to local conditions and that promote greater equality in waste management practices.

3.1 *Japan*

Japan has long been recognized as a world leader in MSW management. The country's waste management system is based on the 3R principle—reduce, reuse, and recycle—and emphasizes the importance of waste reduction at the source. The Japanese government has set ambitious targets for waste reduction and recycling, with a goal of reducing waste by 25% by 2030 and achieving a recycling rate of 60%.

The Japanese government works to educate residents and businesses on the importance of waste reduction, recycling, and proper waste sorting practices. This has led to a high level of public participation in waste reduction and recycling programs, and a cultural shift toward viewing waste as a resource (Mekonnen and

Table 3 Comparison of MSWM indicators and challenges in selected countries

Indicator	Bangladesh	China	Brazil	Switzerland	Israel	USA
Waste generation rate	0.57 kg/cap/day	0.45 kg/cap/day	0.89 kg/cap/day	0.71 kg/cap/day	1.53 kg/cap/day	2.20 kg/cap/day
Collection coverage	56%	100%[a]	70%	–	–	–
Waste treatment methods	Open dumping Landfilling Composting	Landfilling Incineration Composting	Landfilling Composting Open dumping	Incineration Landfilling Composting Anaerobic digestion	Landfilling Incineration Composting	Landfilling Incineration Composting
Recycling rate	~15%	26.7%	~<5%	53%	20%	25.7%
Main MSWM problems	Lack of regulations for disposal Inadequate financial support Municipal authorities' inability to tackle the magnitude of the problem Insufficient budget for both collection and treatment Lack of efficient collection coverage, transport services, treatment, recycling, and disposal facilities	Limited landfill space and inadequate waste disposal infrastructure Insufficient funding Limited public awareness and education Limited enforcement of MSWM regulations and laws	Improper waste disposal Lack of effective education campaigns Week political commitment Inconsistency in statistical data Lack of practical viability of long-term strategies to support recycling	High level of waste generation per capita Increasing trend of MSW generation over the years	Limited land availability for landfilling High population density Illegal dumping Lack of public awareness	Lack of uniform national policy for recycling and funding Many landfills are reaching capacity Low-income and minority communities are disproportionately impacted by MSWM problems
Reference	Alam and Qiao (2020)	Beka and Meng (2021), Ding et al. (2021), Zhu et al. (2021)	Morais et al. (2021), Penteado and de Castro (2021)	Magazzino et al. (2020)	Di Maria et al. (2020)	Di Maria et al. (2020)

MPS Mechanical physical stabilization, *MBT* Mechanical biological treatment, *WtE* Waste to energy
[a] Reported for Shanghai

Tokai 2020). Japan has put in place measures that encourage the reduction of waste and the practice of recycling.

3.2 Germany

Germany is often cited as a world leader in MSW management. The country has a comprehensive waste management system that emphasizes waste reduction, recycling, and energy recovery (Azevedo et al. 2021). In 2018, Germany generated approximately 51 million tons of MSW. Around 64% of this waste was recycled or composted, while 34% was incinerated for energy recovery. Another important aspect of Germany's waste management system is its extended producer responsibility (EPR) program. Under the EPR program, manufacturers and importers of certain products are responsible for the costs associated with their disposal or recycling.

Germany has also introduced measures that encourage the reduction of waste and promote recycling. For example, households are charged for the amount of nonrecyclable waste they generate, which creates a financial incentive for waste reduction and recycling (Azevedo et al. 2021).

3.3 India

It is a rapidly growing country with a population of over 1.3 billion people, and with this growth comes significant challenges in managing MSW. The current state of MSW management in India is characterized by inadequate infrastructure, poor waste collection and disposal practices, and limited awareness among the general public about the importance of proper waste management (Chand Malav et al. 2020).

Another challenge is the lack of effective policies and regulations governing waste management in India. While there are a number of laws and regulations in place, the enforcement of these regulations is weak, and the penalties for noncompliance are not severe enough to act as a deterrent. Additionally, many local municipalities are understaffed and underfunded, further complicating the implementation and enforcement of waste management policies and regulations (Chand Malav et al. 2020). In recent years, the Indian government has made efforts to improve the state of MSW management in the country. As part of this initiative, the government has allocated significant funds to build waste treatment plants and improve waste collection and transportation infrastructure.

3.4 Nigeria

The current state of MSW management in Nigeria is a major concern, with limited infrastructure, inadequate funding, and poor public awareness contributing to widespread environmental pollution and public health risks. Nigeria has a population of over 196 million people, making it the most populous country in Africa, and generates an estimated 32 million tons of waste annually, according to the Federal Ministry of Environment (Ibikunle et al. 2019). There is also a lack of education and outreach programs to promote better waste management practices and encourage behavioral change (Ezeudu et al. 2021).

Despite these challenges, there have been some recent efforts to improve MSW management in Nigeria. For example, the government has launched the National Waste Management Program, which aims to improve waste collection, transportation, and disposal in the country. Additionally, several private companies are now investing in waste-to-energy (WTE) projects, which could provide alternative solutions to the country's waste management challenges.

4 Environmental Impacts of Mismanagement of MSW

The mismanagement of MSW has severe negative impacts on the environment (Fig. 3). Some of the negative impacts of the mismanagement of MSW on the environment are the release of toxic chemicals, contribution to climate change, air

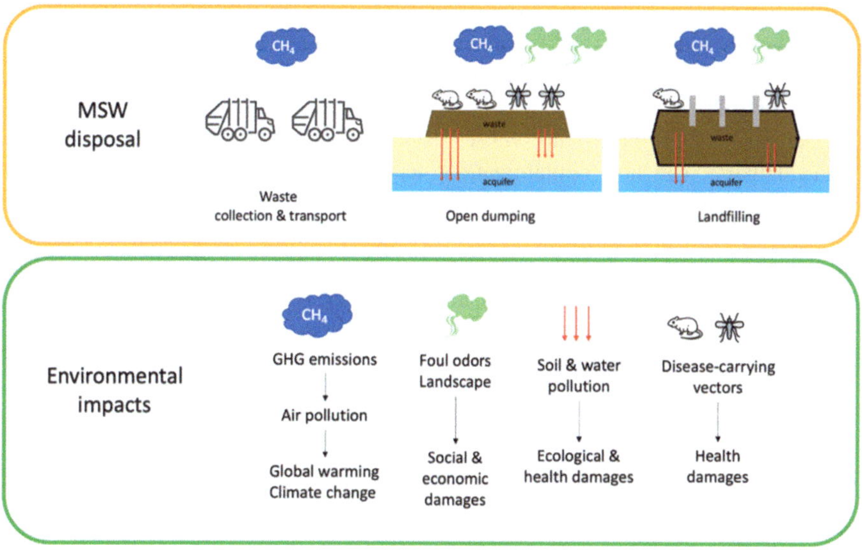

Fig. 3 MSW disposal practices and environmental impacts

pollution, destruction of habitats and biodiversity loss, and economic and social impacts (Kulkarni 2020; Vyas et al. 2022). The improper disposal of MSW can lead to the release of toxic chemicals and pollutants into the environment, which can contaminate soil and water resources. This can lead to health problems for both humans and wildlife, with long-term consequences for biodiversity and ecosystem functioning (Wang et al. 2019).

Additionally, the transportation and processing of waste can also generate carbon emissions. Poorly managed disposal of MSW can lead to the destruction of natural habitats and biodiversity loss. Improper waste disposal can damage ecosystems and harm wildlife, disrupting ecological balance and reducing biodiversity. The accumulation of waste can also change the soil composition, which can negatively affect plant growth and disrupt the food chain.

Finally, the mismanagement of MSW can have negative economic and social impacts. Improper waste management practices can lead to reduced property values, discourage tourism, and increase healthcare costs due to pollution-related illnesses. In addition, the negative effects of MSW mismanagement are often felt most strongly by marginalized communities, who are disproportionately affected by environmental degradation and health risks.

4.1 Water

The leachate from landfills can seep into groundwater and surface water, contaminating them with harmful chemicals and pathogens. Leachate is formed when rainwater percolates through the waste and mixes with organic and inorganic compounds, producing a toxic cocktail of pollutants. These pollutants can include heavy metals, pathogens, and organic chemicals, which can cause a range of health problems, including cancer, neurological disorders, and reproductive problems (Kulkarni 2020).

The disposal of hazardous waste, such as electronic waste, chemicals, and medical waste, can also contaminate water resources. These wastes can contain toxic chemicals that can leach into water sources, leading to pollution and harm to aquatic ecosystems. Marine litter, such as plastic waste and debris, can harm marine life, damage aquatic ecosystems, and lead to water pollution. MSW mismanagement can also contribute to the eutrophication of water bodies. The organic matter in MSW can lead to an increase in nutrient levels in water, which can cause an overgrowth of algae and other aquatic plants.

4.2 Soil

Improper disposal of hazardous waste, such as electronic waste, chemicals, and medical waste brings about the release of toxic chemicals such as heavy metals, polychlorinated biphenyls (PCBs), dioxins, and volatile organic compounds (VOCs) (Vyas et al. 2022). Moreover, the chemicals, such as lead, mercury, and cadmium in the soil, accumulate in the food chain and cause bio-accumulation in the plant eco-system (Rautela et al. 2021).

4.3 Climate Change

The decomposition of organic waste in landfills can lead to the production of methane gas. The United Nations Intergovernmental Panel on Climate Change (IPCC) has estimated that methane emissions from landfills account for about 5% of global GHG emissions, contributing to climate change and its associated impacts. These emissions contribute to climate change and can negatively affect human health (Khan et al. 2022). Transportation and handling of MSW can also contribute to GHG emissions. Collection and transportation of waste to landfills or incinerators require fossil fuel-powered vehicles, which emit GHG.

4.4 Public Health

Open dumping of MSW can lead to the breeding of disease-carrying pests such as rats, flies, and mosquitoes, which can spread diseases such as malaria, dengue fever, and cholera. This is particularly problematic in urban areas, where large amounts of waste can accumulate in densely populated areas (Reddy 2011). These substances are known to be carcinogenic and can cause respiratory problems, skin irritation, and other health problems.

Informal waste pickers, who often work in hazardous conditions without proper protective equipment, are particularly vulnerable to health hazards. They are at risk of injury, respiratory problems, and exposure to hazardous chemicals, which can lead to long-term health problems (Khan et al. 2022).

5 Case Studies Depicting Mismanagement of MSW Caused Significant Environmental Damage

There have been several incidents around the world where the mismanagement of municipal solid waste (MSW) has caused significant environmental damage. Some examples are The Ghazipur landfill in Delhi, India, the Philippines dumping controversy, and the Lago Agrio oil field contamination in Ecuador.

5.1 Ghazipur Landfill in Delhi, India

It is one of the largest landfills in the country, covering an area of over 70 acres and reaching a height of more than 50 meters. It was designed to handle about 1300 tons of waste per day, but it was receiving nearly double that amount. The landfill had been operating for more than 30 years and had exceeded its capacity several times over the years (Ghosh et al. 2019). On September 1, 2017, a part of the landfill collapsed, causing a huge pile of garbage to slide down the slope and crash into a nearby canal. The collapse led to the death of two people and several injuries. The incident also caused significant damage to the environment, including pollution of the nearby water source and damage to the surrounding ecosystem (Law and Ross 2019).

The immediate cause of the collapse was the heavy rainfall that had been occurring in the region, which had made the waste on the landfill heavier and more unstable. However, the root cause of the collapse was the mismanagement of the landfill. The landfill had also been exceeding its capacity for years, causing it to become taller and steeper, which made it more susceptible to collapse (Ghosh et al. 2019). The incident provoked a strong negative response from the public, and the Delhi government was forced to take action. The landfill was closed down, and the government announced plans to build new WTE plants and to increase the recycling of waste. The incident also highlighted the need for better waste management practices in India, including the segregation of waste at source, the promotion of composting and recycling, and the development of sustainable waste management systems (Ghosh et al. 2019).

5.2 Philippines Dumping Controversy

The Philippines dumping controversy, also known as the Canadian waste issue, refers to a dispute between the Philippines and Canada over the dumping of Canadian waste in the Philippines. The controversy began in 2013 when a Canadian company, Chronic Inc., exported 103 shipping containers of mixed garbage labeled as recyclable plastics to the Philippines. The shipment arrived in the Philippines in 2014, and the Bureau of Customs discovered that the containers were filled with household waste, including used adult diapers, food waste, and plastic bottles (Stoett and Omrow 2021).

In response, the Philippine government asked the Canadian government to take back the garbage and threatened to pursue legal action if Canada refused (Stoett and Omrow 2021). The controversy escalated in 2019 when Philippine President Rodrigo Duterte demanded that Canada take back the waste by May 15, 2019, or he would declare war on Canada. Canada initially refused to take back the garbage, citing legal and contractual issues.

5.3 *Bordo Poniente, Mexico*

Bordo Poniente was once the largest landfill in Mexico City, spanning an area of 927 acres. The landfill was in operation for over three decades, accepting over 76 million tons of MSW during this time. However, the mismanagement of the landfill led to significant environmental damage and posed serious health hazards for the residents living in the vicinity (Gutiérrez Galicia et al. 2019). One of the major environmental problems caused by the Bordo Poniente landfill was groundwater contamination. The landfill was located on the outskirts of Mexico City, near the Lerma River, which is a vital source of water for the region. The landfill's lining was inadequate, and the waste was left exposed, which allowed leachate to seep into the groundwater. The contamination of groundwater affected the drinking water supply of the region and posed a significant health hazard to the residents (Gutiérrez Galicia et al. 2019).

The landfill's mismanagement also led to the burning of waste, which caused further air pollution and health hazards (Gutiérrez Galicia et al. 2019). Additionally, the mismanagement of the landfill led to the formation of large amounts of garbage mountains, which posed a significant risk of landslides and slope failures, particularly during the rainy season. The garbage mountains also served as a breeding ground for disease-carrying pests like rats and mosquitoes, posing a health hazard to the residents (Gállego Bravo et al. 2019).

6 Best Practices in MSW Management

MSW management is an important aspect of environmental sustainability. With the increasing amount of waste produced every day, proper management practices are essential to prevent environmental damage. Implementing proper waste management practices is critical in preventing environmental damage. By mitigating pollution, conserving natural resources, mitigating climate change, preserving ecosystems, and protecting human health and safety, the correct handling of MSW can help to create a sustainable environment. Individuals, businesses, and governments must adopt suitable waste management practices to protect the environment and ensure a healthier and sustainable future for generations to come. Most of these practices are consider within the waste hierarchy, which prioritized waste prevention measures over final disposal (Fig. 4). Some of the best practices in MSW management are the following:

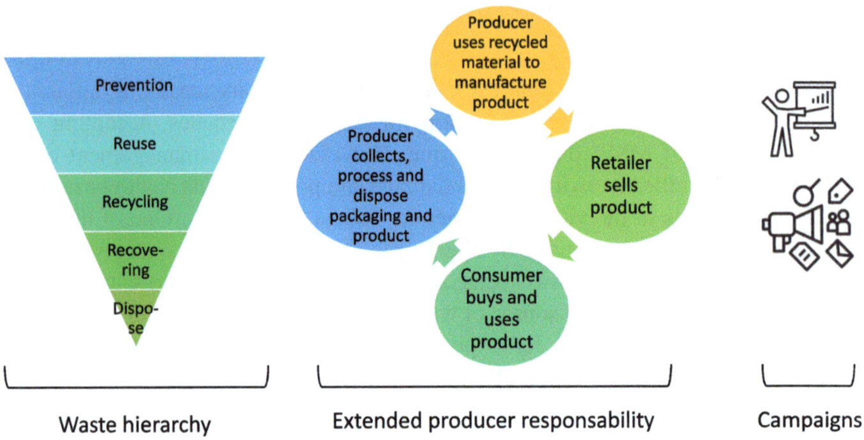

Fig. 4 Best practices of MSW management

6.1 *Waste Reduction and Segregation*

One of the most effective practices in MSW management is waste reduction. This includes strategies such as reducing packaging, reusing items, and recycling. By reducing the amount of waste produced, the overall impact on the environment can be significantly reduced. On the other hand, segregation of waste involves process of separating waste materials into different categories such as organic waste, recyclables, and hazardous waste. This practice enables the different types of waste to be treated and disposed of in the most appropriate manner.

6.2 *Composting*

It is the process in which organic waste is broken down by microorganisms such as bacteria, fungi, and protozoa to produce compost. This compost is rich in nutrients can be utilized for agricultural practices, which will help in increasing the soil fertility. The most common types of aerobic composting methods include windrow composting, aerated static pile composting, and in-vessel composting. Apart from these, compost materials help in preventing soil erosion, soil reclamation, and construction of wetlands and landfill covers. Moreover, this method provides both environmental and economic benefits. It reduces the amount of waste sent to landfills and provides an opportunity to convert organic waste into a valuable resource for agriculture.

6.3 Incineration

Energy can be recovered by the process of incineration or controlled combustion. This can be done through incineration, gasification, or other thermal treatment methods. It provides a source of renewable energy. It is the technology that can be mainly used for MSW, which is containing less moisture after compaction. In this method, solid wastes are heated at a very temperature of 850 °C in the presence of air. There are different types moving grate, fixed grate, rotary-kiln, and fluidized bed are developed to deal with solid waste. However, in this process, carbon dioxide is emitted in the atmosphere and noncombustible materials, i.e., ash is produced. On the other hand, refuse-derived fuel (RDF) (MBT) is the method that can be used for treating combustible fractions of plastic, paper, and cardboard whoch cannot be easily recycled.

6.4 Sanitary Landfilling

Landfills are an important component of MSW management, and proper landfill management practices can significantly reduce their impact on the environment. This includes measures such as proper lining, leachate collection and treatment, and landfill gas management. Sanitary landfills are designed in such way that there is a both landfill gas and leachate collection systems. The collected landfill gas can be used for generation of electricity.

6.5 Polices Intervention and Laws

Extended producer responsibility (EPR): Is an important best practice in MSW management as it holds manufacturers responsible for the environmental impact of their products throughout their lifecycle. By requiring producers to take responsibility for the disposal and recycling of their products, EPR encourages them to design products that are more environmentally friendly and easier to recycle, which reduces waste generation, conserves natural resources, and promotes a circular economy. EPR also incentivizes the development of effective waste management infrastructure and systems, reduces the environmental impact of products, and promotes a more equitable distribution of waste management costs.

7 Case Studies Depicting MSW Best Practices

7.1 San Francisco, USA

San Francisco is often cited as one of the most successful case studies in MSW management in a city. The city has set a remarkable example by reducing its landfill waste by 80% and achieving zero waste in 2020. The success of San Francisco in MSW management can be attributed to several key factors, including a strong political will, a comprehensive regulatory framework, public participation, and the use of innovative technologies (EPA 2021). This commitment is reflected in the city's long-term waste management plan, which sets ambitious goals and targets for reducing waste and increasing recycling and composting. Additionally, the city has launched a number of campaigns to promote waste reduction, such as "Zero Waste Challenge" and "Recycle Myths" (EPA 2021).

Innovation and the use of new technologies have also been crucial to San Francisco's success in MSW management. The city has implemented innovative programs such as the "Pay-As-You-Throw" system, which charges residents and businesses for the amount of waste they generate. This system has been effective in reducing waste and increasing recycling and composting. Additionally, the city has implemented a "Mandatory Recycling and Composting Ordinance," which requires all businesses and residents to recycle and compost. This has helped the city divert more waste from landfills and reduce GHG emissions (EPA 2021).

7.2 Taipei, Taiwan

Taipei, the capital city of Taiwan, is considered a successful case study of MSW management in a city. The city has transformed its waste management practices and infrastructure over the past few decades, making it a model for other cities around the world (Nguyen et al. 2020).

One of the essential factors in Taipei's success in MSW management is public participation. The city has implemented a number of programs and campaigns to encourage residents to reduce waste, recycle, and compost. For example, the city has launched a recycling lottery program, where residents can exchange recyclables for lottery tickets, incentivizing them to recycle more. The city has also implemented a "food waste recycling" program, which requires households and businesses to separate their food waste from other waste and send it to designated facilities for processing.

7.3 Ljubljana, Slovenia

Ljubljana, Slovenia's capital city, has become a benchmark for MSW management worldwide due to its successful transformation of waste management practices and infrastructure over the past few decades. One of the central factors in Ljubljana's success in MSW management is public participation. For example, the city has implemented a "zero waste" campaign, which aims to reduce the amount of waste generated by residents and businesses. The city has also launched a "waste-free market" initiative, where farmers and vendors can sell products without using plastic packaging (Romano et al. 2021).

Innovation and the use of new technologies have also been important to Ljubljana's success in MSW management. The city has implemented several innovative programs, such as the "Ljubljana Bicycle System," which uses bicycles to collect recyclables and compostables from households.

7.4 Adelaide, South Australia

One of the primary reasons for Adelaide's success is its integrated waste management approach, which includes waste reduction, reuse, recycling, and energy recovery. The city has implemented a three-bin system, which includes a green bin for organic waste, a yellow bin for recyclables, and a red bin for residual waste. This system has proven to be effective in diverting waste from landfills, reducing GHG emissions, and promoting sustainable waste management practices (Du et al. 2023).

The city has also implemented several waste management programs to encourage waste reduction and recycling, such as a food waste recycling program and a program to recycle electronic waste. Adelaide has also established a composting facility, which processes organic waste into compost for use in agriculture and landscaping. These programs have not only helped reduce the amount of waste going to landfills but also created employment opportunities and contributed to the local economy (Du et al. 2023). Adelaide has made significant progress toward achieving this goal, with over 80% of waste being diverted from landfills (Arnold et al. 2019).

Adelaide has also adopted sustainable waste management practices in its construction industry, where waste is generated in large quantities. The city requires construction and demolition projects to have waste management plans that promote waste reduction, reuse, and recycling. This has resulted in a significant reduction in construction waste being sent to landfills (Arnold et al. 2019).

8 Conclusions

This chapter provides a comprehensive review of the environmental damages associated with MSW mismanagement, including air and water pollution, soil degradation, and greenhouse gas emissions. The current global waste generation trends show that MSW production is increasing rapidly due to population growth, urbanization, and changes in consumption patterns. The rising waste generation rates have led to the adoption of various waste management methods, including landfilling, incineration, and recycling. However, each method has negative environmental impacts that need to be addressed. The mismanagement of MSW poses significant environmental challenges that require urgent attention. Effective waste management policies and regulations, as well as the adoption of sustainable waste management practices, can help reduce the environmental damages associated with MSW mismanagement. The implementation of the waste hierarchy and circular economy principles can significantly reduce waste generation and promote resource efficiency, leading to a more sustainable and healthier environment for future generations. It is crucial to prioritize sustainable waste management practices to mitigate the adverse impacts of MSW mismanagement and ensure a sustainable future for our planet.

References

Alam O, Qiao X (2020) An in-depth review on municipal solid waste management, treatment and disposal in Bangladesh. Sustain Cities Soc 52:101775. https://doi.org/10.1016/j.scs.2019.101775

Arnold J, Catanzariti A, Chiswell S, Clothier C, Giesecke A, Molly N, Murawsky M (2019) Responsible waste management and the creation of circular economies. https://www.lgprofessionalssa.org.au/resources/LG%20Professionals/Professional%20Development/ELP/Past%20projects/2019%20Final%20Group%20Project%20-%20ELP%20Group%204%20-%20Responsible%20Waste%20Management.pdf

Azevedo BD, Scavarda LF, Caiado RGG, Fuss M (2021) Improving urban household solid waste management in developing countries based on the German experience. Waste Manag 120:772–783. https://doi.org/10.1016/j.wasman.2020.11.001

Beka DD, Meng X-Z (2021) Redesign solid waste collection and transference system for Addis Ababa (Ethiopia) based on the comparison with Shanghai, China. OALib 08(05):1–23. https://doi.org/10.4236/oalib.1107470

Chand Malav L, Yadav KK, Gupta N, Kumar S, Sharma GK, Krishnan S, Rezania S, Kamyab H, Pham QB, Yadav S, Bhattacharyya S, Yadav VK, Bach Q-V (2020) A review on municipal solid waste as a renewable source for waste-to-energy project in India: current practices, challenges, and future opportunities. J Clean Prod 277:123227. https://doi.org/10.1016/j.jclepro.2020.123227

Di Maria F, Mersky RL, Daskal S, Ayalon O, Ghosh SK (2020) Preliminary comparison among recycling rates for developed and developing countries: the case of India, Israel, Italy and USA. In: Sustainable waste management: policies and case studies. Springer, Singapore, pp 1–13. https://doi.org/10.1007/978-981-13-7071-7_1

Ding Y, Zhao J, Liu J-W, Zhou J, Cheng L, Zhao J, Shao Z, Iris Ç, Pan B, Li X, Hu Z-T (2021) A review of China's municipal solid waste (MSW) and comparison with international regions: management and technologies in treatment and resource utilization. J Clean Prod 293:126144. https://doi.org/10.1016/j.jclepro.2021.126144

Du L, Zuo J, Chang R, Zillante G, Li L, Carbone A (2023) Effectiveness of solid waste management policies in Australia: an exploratory study. Environ Impact Assess Rev 98:106966. https://doi.org/10.1016/j.eiar.2022.106966

EC (2018) European platform on life cycle assessment. Joint Research Centre Data Catalogue

EPA (2021) Zero waste case study. United States Environmental Protection Agency, San Francisco. https://www.epa.gov/transforming-waste-tool/zero-waste-case-study-san-francisco

EPA (2022) Paper and paperboard: material-specific data. facts and figures about materials, waste and recyclin. https://www.epa.gov/facts-and-figures-about-materials-waste-and-recycling/paper-and-paperboard-material-specific-data

Ezeudu OB, Agunwamba JC, Ugochukwu UC, Ezeudu TS (2021) Temporal assessment of municipal solid waste management in Nigeria: prospects for circular economy adoption. Rev Environ Health 36(3):327–344. https://doi.org/10.1515/reveh-2020-0084

FAO (2014) Food wastage footprint, full-cost accounting

Gállego Bravo AK, Salcedo Serrano DA, López Jiménez G, Nirmalkar K, Murugesan S, García-Mena J, Gutiérrez Castillo ME, Tovar Gálvez LR (2019) Microbial profile of the leachate from Mexico City's Bordo Poniente composting plant: an inoculum to digest organic waste. Energies 12(12):2343. https://doi.org/10.3390/en12122343

Ghosh P, Shah G, Chandra R, Sahota S, Kumar H, Vijay VK, Thakur IS (2019) Assessment of methane emissions and energy recovery potential from the municipal solid waste landfills of Delhi, India. Bioresour Technol 272:611–615. https://doi.org/10.1016/j.biortech.2018.10.069

Gutiérrez Galicia F, Coria Páez AL, Tejeida Padilla R (2019) A study and factor identification of municipal solid waste management in Mexico City. Sustainability 11(22):6305. https://doi.org/10.3390/su11226305

Ibikunle RA, Titiladunayo IF, Akinnuli BO, Dahunsi SO, Olayanju TMA (2019) Estimation of power generation from municipal solid wastes: a case study of Ilorin metropolis, Nigeria. Energy Rep 5:126–135. https://doi.org/10.1016/j.egyr.2019.01.005

Kaza S, Yao L, Bhada-Tata P, Van Woerden F (2018) What a waste 2.0: a global snapshot of solid waste management to 2050. The World Bank. https://doi.org/10.1596/978-1-4648-1329-0

Khan S, Anjum R, Raza ST, Ahmed Bazai N, Ihtisham M (2022) Technologies for municipal solid waste management: current status, challenges, and future perspectives. Chemosphere 288: 132403. https://doi.org/10.1016/j.chemosphere.2021.132403

Kulkarni BN (2020) Environmental sustainability assessment of land disposal of municipal solid waste generated in Indian cities—a review. Environ Dev 33:100490. https://doi.org/10.1016/j.envdev.2019.100490

Kumar S (2016) Municipal solid waste management in developing countries. CRC Press

Law HJ, Ross DE (2019) International solid waste association's "closing dumpsites" initiative: status of progress. Waste Manag Res 37(6):565–568. https://doi.org/10.1177/0734242X19845755

Magazzino C, Mele M, Schneider N (2020) The relationship between municipal solid waste and greenhouse gas emissions: evidence from Switzerland. Waste Manag 113:508–520. https://doi.org/10.1016/j.wasman.2020.05.033

Mekonnen GB, Tokai A (2020) A historical perspective of municipal solid waste management and recycling system in Japan: learning for developing countries. J Sustain Dev 13(3):85. https://doi.org/10.5539/jsd.v13n3p85

Morais L, Nascimento V, Simões S, Ometto J (2021) Regional distance routes estimation for municipal solid waste disposal, case study São Paulo state, Brazil. Energies 14(13):3964. https://doi.org/10.3390/en14133964

Nanda S, Berruti F (2021) Municipal solid waste management and landfilling technologies: a review. Environ Chem Lett 19(2):1433–1456. https://doi.org/10.1007/s10311-020-01100-y

Nguyen KLP, Chuang YH, Chen HW, Chang CC (2020) Impacts of socioeconomic changes on municipal solid waste characteristics in Taiwan. Resour Conserv Recycl 161:104931. https://doi.org/10.1016/j.resconrec.2020.104931

Penteado CSG, de Castro MAS (2021) Covid-19 effects on municipal solid waste management: what can effectively be done in the Brazilian scenario? Resour Conserv Recycl 164:105152. https://doi.org/10.1016/j.resconrec.2020.105152

Rautela R, Arya S, Vishwakarma S, Lee J, Kim K-H, Kumar S (2021) E-waste management and its effects on the environment and human health. Sci Total Environ 773:145623. https://doi.org/10.1016/j.scitotenv.2021.145623

Reddy PJ (2011) Municipal solid waste management : processing, energy recovery, global examples. CRC Press

Romano G, Marciano C, Fiorelli MS (2021) Best practices in urban solid waste management. Emerald Publishing Limited. https://doi.org/10.1108/9781800438880

Stoett P, Omrow DA (2021) The transnationalization of hazardous waste. In: Spheres of transnational ecoviolence. Springer, pp 73–101. https://doi.org/10.1007/978-3-030-58561-7_3

Tejaswini MSSR, Pathak P, Ramkrishna S, Ganesh PS (2022) A comprehensive review on integrative approach for sustainable management of plastic waste and its associated externalities. Sci Total Environ 825:153973. https://doi.org/10.1016/j.scitotenv.2022.153973

Tubiello FN, Rosenzweig C, Conchedda G, Karl K, Gütschow J, Xueyao P, Obli-Laryea G, Wanner N, Qiu SY, De Barros J, Flammini A, Mencos-Contreras E, Souza L, Quadrelli R, Heiðarsdóttir HH, Benoit P, Hayek M, Sandalow D (2021) Greenhouse gas emissions from food systems: building the evidence base. Environ Res Lett 16(6):065007. https://doi.org/10.1088/1748-9326/ac018e

Vyas S, Prajapati P, Shah AV, Varjani S (2022) Municipal solid waste management: dynamics, risk assessment, ecological influence, advancements, constraints and perspectives. Sci Total Environ 814:152802. https://doi.org/10.1016/j.scitotenv.2021.152802

Wang P, Hu Y, Cheng H (2019) Municipal solid waste (MSW) incineration fly ash as an important source of heavy metal pollution in China. Environ Pollut 252:461–475. https://doi.org/10.1016/j.envpol.2019.04.082

Zhu Y, Zhang Y, Luo D, Chong Z, Li E, Kong X (2021) A review of municipal solid waste in China: characteristics, compositions, influential factors and treatment technologies. Environ Dev Sustain 23(5):6603–6622. https://doi.org/10.1007/s10668-020-00959-9

A Detailed Review on the Environmental Problem and Remediation of Anthropogenic Biomass Waste

Swapan Suman, Dilip Kumar Rajak, Ganesh Kumar, Bijendra Kumar, and Jahir Ahamad Jibran

Abstract Biomass waste management is a transnational, ever-growing dilemma for a healthy environment and its related issues. Growing agricultural and industrial sectors, households, and municipal garbages are the major sources of biowaste. The careless handling of these wastes results in a continuous buildup of toxic contaminants, which have a detrimental effect on the environment and living things. Burning biomass is a substantial source of pollution affecting local, regional, and global air quality, human health, land, and the environment. The high anthropogenic activity also contributes to soil degradation, making soil health and safety a crucial issue. Using organic bio wastes results in the release of macro- and micronutrients, carbon sequestration, and the immobilization and stability of heavy metals. The usage of biowaste increases carbon sequestration, which aids in reducing climate change and global warming. The stimulation of shoot and root length, biomass production, grain yield, chlorophyll content, and a reduction in oxidative stress, soil amendment with biowaste promotes soil activity and plant productivity. This article reviews the sustainable utilization of biomass waste processes and possible future novel practices.

Keywords Biomass waste · Anthropogenic · Carbon sequestration · Climate change

S. Suman (✉)
Department of Mechanical Engineering, Meerut Institute of Engineering and Technology, Meerut, Uttar Pradesh, India
e-mail: swapan.suman@miet.ac.in

D. K. Rajak
Department of Chemical Science & Engineering, Kathmandu University, Kathmandu, Nepal

G. Kumar
Department of Ocean Engineering, Indian Institute of Technology Madras, Chennai, India

B. Kumar
Department of Civil Engineering, Bakhtiyarpur College of Engineering, Patna, India

J. A. Jibran
Department of Mechanical Engineering, Kathmandu University, Dhulikhel, Kavre, Nepal

P. Pathak et al. (eds.), *Anthropogenic Environmental Hazards*,
https://doi.org/10.1007/978-3-031-41013-0_9

1 Introduction

The pattern of energy consumption in any nation can be used to infer that nation's level of socioeconomic development. Natural gas, oil, and coal are examples of fundamental energy sources that are frequently regarded as the most essential energy sources on the planet. According to the most recent statistics, the output of primary energy sources has increased by between 75% and 79% relative to other energy sources (Nasir et al. 2013), and the global consumption of primary energy sources in 2012 reached 12,476.76 million metric tonnes (British Petroleum 2013). However, renewable energy only accounts for between 13 and 18% of the total demand for primary energy. In order to prevent the depletion of inputs from fossil fuels, which are significant energy sources, there has been a strong emphasis on the application of renewable energy as alternative energy option (Halder et al. 2014). There are several forms of energy derived from non-conventional and fossil fuels that can be used as alternatives to these fuels. The terms "alternative sources of energy" and "renewable energy resources" are frequently used interchangeably. Renewable energy resources that utilize domestic resources are able to offer energy services without or nearly nil air pollutant and greenhouse gas emissions. The use of anthropogenic biomass and its application to utilize as alternative energy is causing a lot of harm to the environment. India produces close to 750 million metric tons of solid anthropogenic biomass annually [including straw, bagasse, husks, shells, and agriculture-related residues such as hardwood chips, sawdust, wood bark, etc.] (Pallavi et al. 2014). This anthropogenic biomass consists of fiber, biogases, husks, and shells. Anthropogenic biomass is one of the most advantageous and cost-effective renewable energy resources on the planet (Saratale et al. 2019). Agricultural by-products such as wood dust, wheat husk, and rice husk are also used as basic materials in the paper and pulp manufacturing industries. Due to the immediate effect that burning these wastes would have on the adjacent ecosystem, disposing of large quantities of these wastes is a particularly challenging task. Consequently, utilizing these residues as a renewable fuel is a formidable obstacle for us.

In order to comprehend the effects that the combustion of anthropogenic biomass has on the atmosphere, it is crucial to provide models with consistent parameterizations and dependable uncertainties. In the past decade, numerous publications were published regarding chemical, physical, and thermodynamic properties of the particles produced by the burning of anthropogenic biomass. Their qualities, smoke particles are well-known. Eighty to ninety % of their volume, for instance, is in the accumulation stage. Smoke particles consist of approximately 50–60% organic carbon and 5–10% black carbon. Human-made smoke particles are capable of reflecting and absorbing solar energy. If the updraft velocity is sufficient, smoke particles may serve as cloud condensation nuclei.

2 Biomass: Sustainable Natural Resource

The term "biomass" refers to all biological matter on Earth. These are the various varieties of compounds capable of storing solar energy. Plants are able to continuously produce biomass through a process known as photosynthesis (McKendry 2002; Demirbas 2001).The American Bioenergy Association defines "biomass" as any ligno-cellulosic organic matter primarily composed of carbon, hydrogen, oxygen, and nitrogen. Biomass is comprised of trees, plants, and their by-products, as well as plant fibers, animal wastes, industrial trash, and municipal solid waste (American Bioenergy Association 2005). Biomass is both a renewable and an alternative source of energy. It can be pyrolyzed to produce a liquid product that can be used as a liquid fuel or as a starting feedstock for recovering valuable compounds, and it can be gasified to produce valuable gas. These are only a few of the applications that can use it. Direct combustion can be used to generate heat and power, or it can be converted into a liquid through pyrolysis. The use of biomass has numerous advantages, including the fact that it is an eco-friendly fuel, that it is plentifully obtainable on earth, that it is inexpensive in agriculturally oriented nations, and that it does not generate carbon dioxide.

In 2021, biomass was responsible for the production of around 4835 trillion British thermal units (TBtu), which is equivalent to approximately 4.8 quadrillion Btu and almost 5% of the total primary energy consumption in the United States. About 2316 TBtu of that total came from biofuels, the majority of which were in the form of ethanol. Another 2087 TBtu came from wood and wood-derived biomass, while the remaining 431 TBtu came from the biomass found in municipal solid waste and sewage, animal manure, and agricultural leftovers (Table 1).

Figure 1 shows typical natural photosynthesis in which the product is sugar and oxygen (O_2) by converting the solar energy (Nguyen et al. 2007). The process of photosynthesis allows anthropogenic biomass to absorb CO_2 from the atmosphere. This CO_2 (carbon dioxide) is then burned in a variety of energy-generating technologies. Anthropogenic biomass can be considered the largest renewable energy source currently in use. In 2008, 10% of the world's annual primary energy supply was attributed to anthropogenic biomass (IPCC 2014; Mohan et al. 2006; Zoulalian 2010). Anthropogenic biomass can contribute to global energy demands. The annual global biomass production is estimated to be less than 147 billion metric tons and the combustion of biomass produces no carbon emissions. This means that the same

Table 1 Biomass energy use by consuming sector in 2021

Sl. No.	Sectors	Amount (in TBtu)	% Age shares
1	Industrial	2313	48%
2	Transportation	1477	31%
3	Residential	464	10%
4	Electric power	435	9%
5	Commercial	147	3%

Source: U.S. Energy Information Administration (public domain)

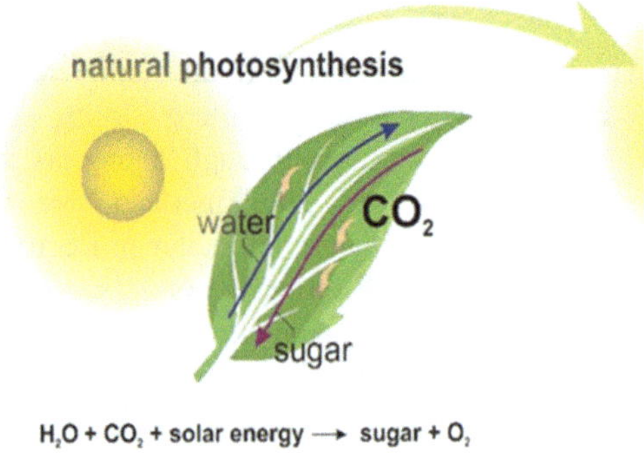

Fig. 1 Photosynthesis (Figure is taken from Nguyen et al. 2007)

quantity of carbon will be released during combustion as was absorbed by plants during their growth. However, experts continue to debate the extent to which the production of biofuels can achieve carbon neutrality.

3 Anthropogenic Biomass in Environmental Problems

The presence of one or more pollutants in the atmosphere in such quantity is harmful for health or wellbeing of humans, animals, or plants. Air pollution is the discharge of hazardous substances into the environment that pollute the air. Air pollution may have detrimental effects not only on people's health but also on the environment and their property. This has contributed to the thinning of the protective ozone layer in the atmosphere leads to climate change. The air quality standard (AQS) is the amount of a pollutant that is permitted to be present outside a structure and that, in order to protect public health, must not exceed the AQS during any given time period. Air pollution is considered to be the presence of hazardous gases and suspended particles in excess of the AQSs. The term air quality criterion (AQC) refers to the varying levels of air pollution and durations of exposure at which certain adverse effects on health and comfort become apparent. According to Morrison and Boyd (1983), Ozone (O_3), carbon monoxide (CO_2), nitrogen dioxide (NO_2), lead (Pb), persistent organic pollutants (POPs), suspended particulate matter, and sulfur dioxide (SO_2) are the most prevalent forms of air pollution. The contribution of CO_2 (Carbon dioxide) to global warming has received increasing attention in recent years. The greenhouse effect is caused by gases with greater heat capacity than oxygen and nitrogen, meaning their molecules contain at least three atoms.

Higher concentrations of greenhouse gases (GHGs) in the atmosphere are predicted to result in a rise in global surface temperature. If anthropogenic biomass fuels are used in an environmentally responsible manner, there will be no overall increase in CO_2 concentration in the atmosphere. Some would even argue that the use of sustainable anthropogenic biomass would result in a net reduction of CO_2 in the atmosphere (Tester et al. 1991). This is predicated that all of the CO_2 released by the combustion of anthropogenic biomass was only absorbed by photosynthesis. Therefore, increasing the use of fuels derived from anthropogenic biomass as opposed to fossil fuels will help reduce the likelihood of global warming, which is caused by increased CO_2 levels in the atmosphere.

The majority of scientists believe that if CO_2 and other gases known as greenhouse gases continue to increase in the atmosphere, then it is likely that the global average temperature will continue to rise (Cline 1992). Over the span of the last century, scientists have recorded around 0.56 °C rise in the global average temperature. The term for this temperature increase is "global warming" or "climate change." Recent warming projections for the next century have been increased to between 1.5 and 5.8 K, based on the assumption that current trends will persist. Changes in the global climate may have a variety of effects, including, but not limited to, rises in sea level, shifts in precipitation patterns, increased weather variability, and other possible outcomes. The production and consumption of energy by humans account for 60% of their impact on the climate. Other human activities, including agriculture (12%), land-use modification (9%), and the use of compounds such as chlorofluorocarbons (15%), all contribute to the increase in greenhouse gas concentration in the atmosphere. The remaining 4% increase is attributable to additional human activities. At present, the annual rate of GHGs emissions is increasing. This trend is anticipated to persist. Due to the issue of global climate disruption, energy concerns will be the focus of a substantial quantity of political action on a global scale. This is a topic that raises essential questions regarding politically sensitive issues such as the sovereignty of individual nations and global justice. The Kyoto conference, which took place in 1997 and was attended by 160 nations, and other global summits, such as the Rio and Montreal meetings, have demonstrated that the majority of governments recognize the need to resolve the issue. The temperature of the earth's surface has increased by approximately 0.6 K over the past century, and as a direct consequence, the sea level is predicted to have risen by approximately 20 cm. If the atmospheric concentrations of greenhouse gases, which are primarily caused by the combustion of fossil fuels, continue to increase at the current rates, the global temperature may increase by 2–4 K over the next century. This is consistent with predictions. According to this prognosis, the sea level could rise between 30 and 60 cm before the turn of the century (Colonbo 1992).

The US Environmental Protection Agency (EPA) lists carbon monoxide, lead, nitrogen dioxide, ozone, suspended particulate matter, and sulfur dioxide as important air pollutants. Human activities including burning coal, oil, and natural gas emit CO_2, the main greenhouse gas. They warm the world. The greenhouse effect, which is affected by climate change, is affected by fossil and anthropogenic biomass fuel emissions of CO_2. 54% of CO_2 emissions come from OECD countries. Inefficient burning of carbon-based fuels like petrol, diesel, and biomass produces CO. This produces odorless, colorless carbon monoxide. It may also be created by burning natural and man-made things like tobacco. It reduces blood oxygenation. It may impair reaction time and tyre us. CO AQS averages 2.0–4.0 mg/m^3. The sun's UV light creates ozone (O_3) in the upper atmosphere. This vital atmosphere screens the

Table 2 Effects of air pollution on the health of infants

Air pollutants	Exposure concentration	Effect	Reference
SO_x		Infant deaths	Bobak and Leon (1992)
NO_x		Respiratory system symptoms, asthma deteriorations, reduced pulmonary function development	Lipsett et al. (1997), Shima and Adachi (2000), Gauderman et al. (2000)
PM_{10}	≥250 ppb	Enhanced allergen reactivity, infant mortality primarily due to respiratory illness, sudden neonatal death syndrome	Woodruf et al. (1997), Strand et al. (1998)
O_3	>110 ppb	37% increase in asthma emergency department appointments, asthma incidence increased threefold in the group with high ambient ozone and high outdoor activity	Tolbert et al. (2000), McConnell et al. (2002)
	>120 ppm	Decreased FEV1 in asthmatic subjects exposed to allergens, incidence of dyspnea increased by 35 percent per 50-ppb increase	Molfino et al. (1992), Petroeschevsky et al. (2001)

earth from damaging UV rays from the sun. At ground level, this pollution is particularly dangerous. Pollution damages the ozone layer. The stratospheric ozone layer shields earth from the sun's UV radiation. CFCs from aerosol cans, cooling systems, and refrigeration equipment destroy the ozone layer. This results in the formation of 'holes' in the ozone layer, making it simpler for radiation to reach the planet's surface. NO_x emissions are responsible for both smog and acid rain. Its production requires the combustion of hydrocarbons such as gasoline, diesel, and coal.

Children exposed to nitrogen oxides during the winter months may have an increased risk of developing respiratory disorders. The typical adverse effect on the children health by air pollutants is shown in Table 2. SO_x, NO_x, PM_{10} (particulate matter), and O_3 have severe health effects on the infant which are responsible for asthma, enhancing allergen reactivity and death (Trasande and Thurston 2005).

The NO_x air quality standards (AQS) limit value falls between 60 and 80 mg/m^3. Carbon dioxide (CO_2) is produced by the incomplete combustion of anthropogenic biomass, along with carbon monoxide (CO), nitrous oxide (N_2O), hydrocarbons (HCs), and particulate matter. It has been demonstrated that smoke from inefficient wood-burning furnaces is one of the most significant risk factors for a variety of health conditions. Using anthropogenic biomass fuels in direct combustion, gasification, or pyrolysis systems could have significant environmental benefits, despite the fact that there are still a number of unanswered concerns regarding these processes. Typically, the output of sulfur dioxide (SO_2), carbon dioxide (CO_2), and waste from anthropogenic biomass power systems will be significantly lower than that of coal combustion and conversion systems. Sulfur dioxide (SO_2) gas produced by the combustion of coal, typically in thermal power facilities sulfur dioxide (SO_2) is a by-product of a number of manufacturing processes, including the production of paper and metals. Significantly contributes to the formation of pollution and acid rain. Sulfur dioxide may cause pulmonary conditions. The AQS value of SO_2 in the air ranges from 60 to 80 mg/m^3. CFCs, also known as chlorofluorocarbons, are gases emitted primarily by refrigeration and air-conditioning systems. When CFCs are discharged into the atmosphere, they rapidly reach the stratosphere. They deplete the earth's ozone layer by interacting with few other gases.

A hazardous air pollutant (HAP) is an air pollutant that endangers human health or the environment due to its ambient concentrations, bio-accumulation, or depositing over time. This contaminant can be inhaled and otherwise ingested. The flue gas was tested for harmful chlorinated hydrocarbons (polychlorinated biphenyls, benzenes, dioxins, and furans) and poly-aromatic hydrocarbons (Ruokojarvi et al. 2000). Lavric et al. (2004) measured dioxins in wood combustion gases and particulates. Dioxins persist and bioaccumulate. Dioxins are persistent organic pollutants (POPs) because they may travel vast distances. Wood emits less than fossil fuels because it has less sulfur and nitrogen. Thus, wood emits almost no sulfur oxide (SO_x) and less nitrogen oxide (NO_x) when the temperature is managed to avoid nitrogen oxidation from the ambient air (McIlveen-Wright et al. 2001).

3.1 Greenhouse Gas Impact

The increased levels of greenhouse gases (GHGs) in the atmosphere are leading to an increase in the average surface temperature of the globe. Carbon dioxide (CO2) is the most important greenhouse gas in terms of its contribution to the warming of the planet. Coal is responsible for between 30 and 40 percent of the world's carbon dioxide (CO_2) emissions that are created by fossil fuels at the time. Nearly all carbon emissions come from the burning of fossil fuels like coal, oil, and natural gas. This accounts for around 98% of all carbon emissions. The term "greenhouse effect" refers to the phenomenon in which there is an overall rise in temperature on Earth. This is due to the presence of a number of gases in the atmosphere, including carbon dioxide (CO_2), oxides of nitrogen (NO_x), methane (CH_4), and other gases (Demirbas 2005).

Table 3 shows the typical emission of GHGs from the combustion of diesel, coal and natural gases. The CO_2 emission from each of the fissile fuel is maximum whereas the emission of CH_4 is always remains $\geq$0.5% (Karmaker et al. 2020).

3.2 Pollution of the Atmosphere

The term "air pollution" refers to a broad category that includes a large number of distinct types of pollution. The presence of one or more pollutants in the atmosphere in such quantity and duration that their presence is harmful or has the potential to be

Table 3 Typical emission of GHGs from the combustion of fossil fuels (Karmaker 2020)

Type of fuel	Emission parameters	Emission (kg/kWh)
Diesel	CO_2	0.76
	CO	0.01
	Unburned hydrocarbons	0.0002
	Particulate matters	0.00003
	SO_2	0.002
	NO_x	0.005
Coal	CO_2	0.90
	SO_2	0.007
	NO_x	0.004
Natural gas	CO_2	0.566
	CO	0.0018
	Particulate matter	0.0000525
	NO_x	0.0039

harmful to the health or well-being of humans, animals, or plants It is the discharge of harmful compounds into the environment that pollutes the air. Air pollution may have negative effects on people's health, the environment, and their property. As a consequence, the protective ozone layer in the atmosphere has become thinner, which contributes to climate change. The personal automobile is the most significant contributor to urban air pollution on a global scale. The emissions from the millions of vehicles and trucks on the planet's roads constitute a global problem. However, the combustion process of a petroleum-based engine releases other compounds into the atmosphere in addition to carbon dioxide and water. Many of these molecules are either smog-producing or openly hazardous, and some of these substances are both. This contains numerous lead compounds that are produced by an engine that runs on leaded petrol (Morrison and Boyd 1983). It also includes unburned hydrocarbons, carbon monoxide, and nitrogen oxides. Both the petroleum and automotive industries are now going through a period of transition. The use of anthropogenic biomass fuels in various thermochemical process may have significant positive effects on the environment, despite the fact that many questions remain about these processes. Carbon dioxide (CO_2), sulfur dioxide (SO_2), and ash emissions would typically be substantially lower in anthropogenic biomass power systems than in coal combustion and conversion systems (Bain et al. 1998).

In addition, gases that are acidic in nature, such as HCl, as well as heavy metals, such as lead, may be released into the atmosphere. According to Donovan Associates' research from 1995, the by-products of incomplete combustion include carbon monoxide (CO), hydrocarbons (HC), volatile organic compounds (VOC), and polycyclic aromatic hydrocarbons (PAH). The thermal pyrolysis of a particular type of anthropogenic biomass generates smoke when there is insufficient oxygen to prevent combustion but sufficient oxygen to permit detrimental reactions. A recent analysis of the components of anthropogenic biomass smoke revealed that it contains approximately 400 volatile components, including acids, alcohols, carbonyls, esters, furans, lactones, phenols, polycyclic aromatic hydrocarbons (PAHs), and various other compounds. These components were extracted by distillation (Luthe et al. 1997). A study was conducted on the dioxin emissions generated by power boilers fueled with salt-laden wood waste. When combustion conditions were favorable, it was discovered that controlling dioxin emissions could facilitate effective control (stack dioxin emissions of 0.064–0.086 ng/m^3). It has been suggested that the combustion of anthropogenic biomass, specifically forest fires, is a significant source of dioxins and furans in the environment.

3.3 Acid Rain

It is common knowledge that the majority of energy-related activities are the primary causes of acid precipitation. Acid rain is a term used to characterize a variety of processes that are more accurately termed acidic deposition. Acid rain is a term used to characterize a number of different processes. When petrol, diesel, and coal are burned, sulfur dioxide and nitrogen oxides are released into the atmosphere. These compounds may react with atmospheric water vapor to produce acid rain, which can then descend to the ground as rain, snow, or fog. Natural sources, such as volcanoes, are also capable of releasing these gases into the atmosphere. NOx emissions are responsible for both smog and acid rain. One of the primary causes of air pollution is the combustion of gasoline, diesel, and coal in residential and commercial settings, such as automobiles and power plants. The average air quality standards (AQS) value of suspended particulate matter (SPM) in respirable air is between 140 and 200 mg/m^3 (Demirbas 2005). Acid rain is detrimental not only to human health but also to water, soil, and forest resources. It is rumoured to be corrosive to structures and hazardous to human health. International agreements have been made to restrict the amount of sulfur and nitrogen oxide emissions, as this issue impacts nations worldwide. These agreements have been executed.

4 Anthropogenic Biomass in Environmental Remediation

4.1 Biosorption and Bioaccumulation

Depending on the type of central agent utilized in the remediation process, abiotic and biotic approaches can be distinguished from one another. These methodologies serve as the foundation for the elimination of hazardous substances. Typically, the application of a biotic (physico-chemical) methods, such as membrane filtration, ion exchange, adsorption and absorption, precipitation, and oxidation, depends on the concentration of the pollutant being treated. This is due to the fact that some of these procedures can harm the environment. Abiotic methods include membrane filtration, ion exchange, adsorption and absorption, and precipitation (Crini 2006). As a result, research in recent years has focused on biological approaches, which have demonstrated several advantages, including being environmentally benign, cost-effective, and able to be conducted in situ while maintaining a high level of efficiency. Particularly, biosorption and bioaccumulation indicated that there is a significant potential for the elimination of hazardous substances, especially pigments and heavy metals. This article provides a thorough analysis of the numerous types of biomass that can be utilized in biosorption and bioaccumulation processes. This article's primary objective is to provide a summary of recent findings regarding the use of both living and nonliving biomass in the process of cleaning up polluted environments.

Although biomass is utilized in both biosorption and bioaccumulation, the mechanisms involved and the levels of efficiency attained are distinct. During the bioaccumulation process, living cells absorb pollutants across the cell membrane and accumulate them within the cells as a consequence of the cells' metabolic cycle (Malik 2004). Biosorption processes, on the other hand, are exclusive to dead anthropogenic biomass and are commonly regarded as a passive mechanism for the retention of contaminants on the cell wall. This is contingent on the type of anthropogenic biomass (Vijayaraghavan and Yun 2008). There are two possible explanations for how living and decomposing anthropogenic biomass absorb metal ions. One of the models is the attachment of metal ions to the cell wall surface (extracellular). This type of metal adsorption is shared by both living and decomposing anthropogenic biomass. The second model proposes that metals infiltrate cells through an active uptake process, also known as bioaccumulation. This occurs when a contaminant passes through the cell membrane and accumulates in living cells. In any event, both living and nonliving anthropogenic biomass possess the ability to absorb metals. The biosorption of metals by anthropogenic biomass is primarily determined by the components present in the cell, especially those that are accessible via the cell surface and the spatial organization of the cell wall. According to Volesky (2007), the most significant of these groups are carbonyl (ketone), carboxyl, sulfhydryl (thiol), sulfonate, thioether, amine, secondary amine, amide, imine, imidazole, phosphonate, and phosphodiester. Other groups included in this category include phosphonate and phosphodiester. It is common knowledge that the use of living anthropogenic biomass as a biotreatment solution cannot be considered for the continuous treatment of highly toxic contaminants. This is due to the fact that the accumulated amount of toxicant will ultimately reach saturation, at which point the cell culture will have died (Eccles 1999). In contrast, deceased or inactive anthropogenic biomass is more adaptable to the concentration of toxicants and environmental conditions, and as a result, it has numerous advantages over living species. In addition, anthropogenic biomass derived from living organisms may require a nutrient supply in order to carry out their metabolic processes. Because of this, either the biological oxygen demand (BOD) or the chemical oxygen demand (COD) of the effluent or the solution may increase (Hemambika et al. 2011). Due to these additions, the approach that employs active anthropogenic biomass is associated with higher costs.

4.2 Captured Carbon Dioxide

Recent global interest has increased in finding "carbon efficient" waste management solutions that conserve energy, reduce carbon dioxide (CO_2) emissions, and make productive use of waste products. It is well known that industrialized nations have pledged to reduce atmospheric carbon dioxide (CO_2) emissions by implementing carbon capture and storage initiatives; however, their ability to fulfil this commitment is limited due to the high costs involved and the unreadiness of the underlying technology. Carbon dioxide utilization (CCU) is the process of using carbon dioxide (CO_2) as a feedstock to manufacture a variety of products, such as construction materials, polymers, and fuels (Armstrong and Styring 2015). Recent technological advancements have created opportunities for CCU. To keep costs low, however, the successful full deployment of CCU technology will depend in part on the direct use of point emissions of carbon monoxide (CO) and carbon dioxide (CO_2) or, if necessary, their preferential capture through the use of, for instance, low-cost sorbents, which may even be waste-derived (Kaithwas et al. 2012). This is due to the fact that direct use of carbon monoxide (CO) and carbon dioxide (CO_2) point emissions will enable for the most efficient use of the technology (Belmabkhout et al. 2011). Carbon dioxide (CO_2) conditioning of cementitious materials has been a widespread practice for decades. This technique has been employed, for instance, for the rapid hardening of calcium silicate-based materials (Berger et al. 1972) and concrete objects such as roofing tiles (Maries and Hills 1986). Carbonation has been demonstrated to be effective for stabilizing soil contaminated with a variety of heavy metals and solidifying cement-based waste forms (Lange et al. 1996, 1997). Zinc, copper, and lead are examples of these metals (Whitehead et al. 2003). Carbonation as a means of reducing risk and producing engineered materials lends credence to the notion that the controlled carbonation of residues derived from anthropogenic biomass may be advantageous. Annual volumes of Gt of anthropogenic biomass by-products are not managed sustainably. The disposal of anthropogenic biomass wastes ought to adhere to a "hierarchy of waste management" that places energy recovery and disposal at the bottom of the list of desired alternatives if the wastes have the potential for other applications, as they do. According to the hierarchy of waste management alternatives presented by the United Nations Environment Programme (2011), energy recovery and disposal are the two least desired options. In spite of this, the use of anthropogenic biomass produced in a sustainable manner over the long term as a substitute feedstock for carbon-intensive industries and fossil fuels results in more persistent CO2 reductions than preservation. According to Searle and Malins (2013), the EU Trash Framework Directive requires that measures be taken in order to decrease the amount of garbage produced to the maximum degree practicable, lessen dependency on landfills, and enhance recycling rates. It has been mandated by the Department of Energy (DOE) and the Department of Agriculture (USDA) of the United States of America that by 2022, 5% of heat and power energy, 20% of liquid transportation fuel, and 25% of chemicals and materials must be derived from anthropogenic biomass (Perlack 2005; Balan 2014). These percentages are to be met in order to comply with the mandates. It is well known that the world's anthropogenic biomass has a considerable untapped potential for the recovery of both materials and energy (UNEP 2009), but innovative technology that can effectively manage waste anthropogenic biomass is still unavailable. This becomes even more significant when one considers the possibility of utilizing refuse products to alleviate pressure on virgin material resources (such as soil and natural aggregates).

4.3 *Utilization as Biochar for Waste Management and Disposal*

According to Matteson and Jenkins (2007), agricultural and animal wastes have the potential to present severe environmental concerns, which might ultimately result in the pollution of ground and surface rivers. This garbage, together with other left-overs, may be utilized as fuel for pyrolysis bioenergy, which is a success for the environment as well as the economy (Bridgwater 2003). According to Matteson and Jenkins (2007), several waste streams hold the economic potential for energy recovery. This is particularly true in places that have a stable supply of feedstock. According to Kwapinski et al. (2010), this results in a reduction of the amount of energy that is required for transportation as well as the emissions of methane (CH4) that would be created if these materials were disposed of in landfills. Hossain et al. (2011) noted that pyrolysis provides the opportunity to convert waste material into biochar. This would lead to an improvement in waste management, a decrease in the costs associated with transporting trash, and a decrease in the quantity of waste produced. As a result of the disposal of organic waste in landfills and the anaerobic digestion of animal manure, it is possible for significant quantities of methane (CH_4) and nitrogen oxide (N_2O) to be discharged into the atmosphere. Consequently, utilizing this waste for the synthesis of biochar and its applications is an effective waste management strategy that helps to reduce greenhouse gas emissions and the costs associated with conventional waste disposal methods (Kwapinski et al. 2010).

China's wastewater treatment facilities generate 25 million metric tons (80% by weight) of sewage sludge annually, which considerably contributes to the country's severe environmental problems. Lu et al. (2012) pyrolyzed waste sewage sediment from these wastewater treatment facilities in order to manufacture biochar for lead absorption. In addition to removing pathogens and reducing biosolids, the production of biochar from sewage sludge converts this organic matter into bio-oil and biochar (Domínguez et al. 2006; Lu et al. 2012). In slow pyrolysis facilities, exceedingly low-grade anthropogenic biomass feedstocks or even by-product garbage can be used to create energy and biochar (Downie 2007). This shifts the carbon in these materials from short-term to long-term carbon cycles. After that, it is possible to recycle the biochar back into the soil, creating a positive feedback loop that will increase crop yields in consecutive cycles (Downie et al. 2011).

Sánchez-García (2015) utilized biochar that had been co-composted with poultry manure and observed the impact of biochar on the degradability of organic matter and the mineralization of nitrogen. This study demonstrated that the addition of biochar to the composting process at a rate of three % by dry weight of poultry manure led to an increase in organic matter (OM) degradability over a shorter composting period, as well as a decrease in the formation of clumps and an increase in the mineralization of nitrogen.

Hydrothermal carbonization (HTC) produces greater biochar yields (50–80%) and lower gas outputs (2%–5%) than dry pyrolysis during the conversion of moist manure or anthropogenic biomass into high-energy fuel and fertilizer. Temperatures between 180 and 300 degrees Celsius are used during this procedure. In addition to charcoal fuel with a high nitrogen-phosphorus-potassium (NPK) content, high thermal carbonization (HTC) provides access to fatty acids and heavy metals of significant economic consequence. The thermal carbonization (HTC) treatment of bovine manure at temperatures between 180 and 260 °C for 5 to 30 min increased the high heat value (HHV) of the ash-free biochar products from 19.1 to 22.1 MJ kg^{-1}. In addition, the nitrogen-phosphorus-potassium (NPK) ratios of the products were 20:20:3, and the solid biochar absorbed phosphorus and other minerals in considerable quantities (Toufiq Reza et al. 2016). Chicken litter was hydrothermally carbonized at 250 °C to create biochar with an HHV of 24.4 MJ kg^{-1}, a nutrient-rich liquid, and a CO_2-dominated vapor (Mau 2016). Ghanim (2016) created poultry litter biochar. Its HVV was 25.17 MJ kg^{-1} 697. The hydrothermal conversion of wheat straw digestate (Reza et al. 2015) and (Suwelack 2016) shows that agricultural and animal waste can be converted into biochar fuel in a controlled environment to produce high-carbon fuels with HHVs comparable to sub-bituminous coal and fertilizer value. This produces high-carbon fuels with sub-bituminous coal-like HHVs. This can provide high-grade fuels with HHVs comparable to sub-bituminous coal.

Conesa et al. (2009) conducted a study to determine how much pollution is produced by the combustion of sewage sludge, cotton textiles, meat and bone meals (MBM), olive pomace, and paper refuse. At 850 °C, pyrolysis produced polyaromatic hydrocarbons (PAHs), the most significant of which was naphthalene. Whenever pyrolysis is performed, these PAHs must be considered. As the temperature increased, the production of volatile hydrocarbons and semivolatile compounds decreased. Light hydrocarbons, semivolatiles, and both monoaromatic and polyaromatic hydrocarbons were produced at intermediate temperatures. The procedure must be carried out in a controlled environment so as not to harm the environment (Conesa et al. 2009). Using these procedures, aromatic compounds can be synthesized. These compounds serve as building blocks for the production of dyes and other chemicals.

4.4 Utilization of Biochar for the Treatment of Wastewater

A water and wastewater purification system that utilizes biochar to remove organic and heavy metal contaminants from water is a relatively new concept with a great deal of potential as a method of removing water contaminants. Agricultural residue feedstocks are composed of cellulose, hemicelluloses, lipids, carbohydrates, and proteins, all of which contain functional groups that can be physically activated by pyrolysis or additional steam and CO_2 treatment to make them more effective at absorbing contaminants. Sugars and proteins are also present in agricultural residue feedstocks (Inyang et al. 2011). Agricultural residue feedstocks contain a multitude of functional groups, each of which can be physically activated through pyrolysis, additional treatment with steam or carbon dioxide, or both. According to research conducted by Yenisoy-Karakaş et al. (2004), the adsorption capacity of biochar is not only determined by its type but also by the concentration of functional groups present on its surface. As a consequence of this, the removal of adsorbates may be better understood. Due to its carbon-structured matrix, high porosity, large surface area, and strong affinity for nonpolar substances such as PAHs, dioxins, furans, and other compounds, biochar is an essential surface sorbent for controlling environmental contaminants (Chen et al. 2011; Regmi et al. 2012; Jiang et al. 2012; Shrestha et al. 2010).

Biochar's superabsorbent properties can be used to remove both organic and inorganic contaminants from soil and water. Its activity is comparable to that of activated carbon (AC), but the two materials differ in their feedstocks, manufacturing processes, and final physiochemical properties. Cao et al. (2009), Lu et al. (2012). X discovered that biochar was superior to commercial AC (Pb) at removing lead. Because so many diverse materials can be used to create biochar, it may be less expensive to use biochar as a Pb adsorption remediation technology than AC (Shang et al. 2012). Aluminum (Al) and manganese (Mn) in acid soils and arsenic (As), cadmium (Cd), copper (Cu), nickel (Ni), and lead (Pb) in heavy metal-contaminated soils can be taken up by biochar because its surface can have a large number of chemically active groups, including OH, COOH, and ketones, among others (Uchimiya et al. 2010; Xu et al. 2011). This means that biochar has a tremendous possibility because biochar is created by separating oxygen-containing functional groups, the majority of biochars have net negative surface charges. As a result, they can be used to economically remove organic pollutants and heavy metal cations from water (Uchimiya et al. 2010; Xu et al. 2011; Qiu et al. 2008; Mohan et al. 2007; Wang and Xing 2007).

Numerous studies (Wang et al. 2010; Chen et al. 2011; Nguyen et al. 2007) have investigated the possibility of removing organic pollutants with biochar produced from plant anthropogenic biomasses. In contrast to AC, however, biochar's ability to remove heavy metals from aqueous solutions is poorly understood (Uchimiya et al. 2010; Tong et al. 2011).

Several additional studies have demonstrated that biochar produced from rice husk, maize straw, peanut straw, olive pomace, and oak wood and bark may be an effective method for removing heavy metals (Wang and Xing 2007; Tong et al. 2011; Liu and Zhang 2009; Pellera et al. 2012; Xue et al. 2012).

5 Conclusions

This literature review focuses on the environmental problem and remediation of biodegradable waste generated by humans. This analysis examines the effects on India's air quality, health, and climate. Smoke particles are a significant source of black carbon and brown carbon, which have negative health and climate effects after the combustion of anthropogenic biomass waste. That contains numerous carcinogenic compounds, including PAHs, hydroxylated, and oxygenated derivatives. It may contribute to the slowing of climate change and the reduction of acid rain, soil erosion, and water pollution through enhanced management. It is believed that the technologies for processing biogenic raw materials for the production of intermediate and final products on integrated and circular platforms will make the coproduction of bio-based products and energy feasible. Biorefineries, which are multistep, multiproduct facilities designed for the eco-efficient production of specific biosourced feedstocks, are viewed as the precursor to widespread applications. It is anticipated that these applications will evolve over time.

Acknowledgments The work was supported by the Department of Chemical Engineering, Meerut Institute of Engineering and Technology, Meerut (UP).

References

American Bioenergy Association (2005) What is biomass [on-line]. Available: www.Biomass.org/index_files/page0001.htm

Armstrong K, Styring (2015) Assessing the potential of utilization and storage strategies for post-combustion CO_2 emissions reduction. Front Energy Res. https://doi.org/10.3389/fenrg.2015.00008

Bain RL, Overend RP, Craig KR (1998) Biomass-fired power generation. Fuel Process Technol 54:1–16

Balan V (2014) Current challenges in commercially producing biofuels from lignocellulosic biomass. ISRN Biotechnol 2014:31. Article ID 463074, http://www.hindawi.com/journals/isrn/2014/463074/

Belmabkhout Y, Serna-Guerrerom R, Sayari A (2011) Adsorption of CO_2-containing gas mixtures over amine-bearing pore-expanded MCM-41 silica application for CO_2 separation. Adsorption 17:395–401

Berger RL, Young JF, Leung K (1972) Acceleration of hydration of calcium silicates by carbon dioxide treatment. Nat Phys Sci 240:16–18

British Petroleum (BP) (2013) Statistical review of world energy, British, http://www.bp.com/en/global/corporate/aboutbp/energy-economics/statistical-review-of-world-energy-2013.html

Bridgwater AV (2003) Renewable fuels and chemicals by thermal processing of biomass. Chem Eng J 91(2–3):87–102

Bobak M, Leon DA (1992) Air pollution and infant mortality in the Czech Republic, 1986–88. Lancet 340:1010–1014
Cao X, Ma L, Bin G, Harris W (2009) Dairy-manure derived biochar effectively sorbs lead and atrazine. Environ Sci Technol 43(9):3285–3291
Crini G (2006) Non-conventional low-cost adsorbents for dye removal: a review. Bioresour Technol 97:1061–1085
Cline W (1992) The economics of global warming. Institute for International Economics, Washington, DC
Colonbo U (1992) Development and the global environment. In: Hollander JM (ed) The energy–environment connection. Island Press, Washington, pp 3–14
Conesa J, Font A, Fullana A, Martín-Gullón I, Aracil I, Gálvez A, Moltó J, Gómez-Rico M (2009) Comparison between emissions from the pyrolysis and 1318 combustion of different wastes. J Anal Appl Pyrolysis 84(1):95–102
Chen X, Chen G, Chen L, Chen Y, Lehmann J, Mcbride MB, Hay AG (2011) Adsorption of copper and zinc by biochars produced from pyrolysis of hardwood and corn straw in aqueous solution. Bioresour Technol 102(19):8877–8884
Demirbas A (2001) Biomass resource facilities and biomass conversion processing for fuels and chemicals. Energ Conver Manage 42:1357–1378
Demirbas A (2005) Potential applications of renewable energy sources, biomass combustion problems in boiler power systems and combustion related environmental issues. Prog Energy Combust Sci 31:171–192
Domínguez A, Menéndez JA, Inguanzo M, Pís JJ (2006) Production of bio-fuels by high temperature pyrolysis of sewage sludge using conventional and microwave heating. Bioresour Technol 97(10):1185–1193. https://doi.org/10.1016/j.biortech.2005.05.011
Downie A (2007) Slow pyrolysis: Australian demonstration plant successful on multi-feedstocks.- In: Bioenergy conference, Finland
Downie A, Zwieten V, Zwieten LV, Smernik RJ, Munroe PR (2011) Terra Preta Australis: reassessing the carbon storage capacity of temperate soils. Agric Ecosyst Environ 140(1–2):137–147. https://doi.org/10.1016/j.agee.2010.11.020
Eccles H (1999) Treatment of metal-contaminated wastes: why select a biological process? Trends Biotechnol 17:462–465
Gauderman WJ, McConnell R, Gilliland F, London S, Thomas D, Avol E, Vora H, Berhane K, Rappaport EB, Lurmann F, Margolis HG, Peters J (2000) Association between air pollution and lung function growth in southern California children. Am J Respir Crit Care Med 162:1383–1390
Ghanim BM (2016) Hydrothermal carbonisation of poultry litter: effects of treatment temperature and residence time on yields and chemical properties of hydrochars. Bioresour Technol 216: 373–380
Halder PK, Paul N, Beg MRA (2014) Assessment of biomass energy resources and related technologies practice in Bangladesh. Renew Sustain Energy Rev 39:444–460
Hemambika B, Rani MJ, Kannan VR (2011) Biosorption of heavy metals by immobilized and dead fungal cells: a comparative assessment. J Ecol Nat Environ 3:168–175
Hossain MK (2011) Influence of pyrolysis temperature on production and nutrient properties of wastewater sludge biochar. J Environ Manage 92(1):223–228
Hossain MK, Strezov V, Chan KY, Ziolkowski A, Nelson PF (2011) Influence of pyrolysis temperature on production and nutrient properties of wastewater sludge biochar. J Environ Manag 92(1):223–228
IPCC (2014) Climate change assessments review of the processes and procedures of the IPCC. Inter Academy Council, The Netherland. https://www.ipcc.ch/pdf/IAC_report/IAC%20Report.pdf
Inyang M et al (2011) Enhanced lead sorption by biochar derived from anaerobically 1320 digested sugarcane bagasse. Sep Sci Technol 6(12):1950–1956
Jiang TY, Jiang J, Xu R-K, Li Z (2012) Adsorption of Pb (II) on variable charge soils amended with rice-straw derived biochar. Chemosphere 89(3):249–256

Kaithwas A, Prasad M, Kulshreshtha A, Verma S (2012) Industrial wastes derived solid adsorbents for CO_2 capture: a mini review. Chem Eng Res Des 90:1632–1164

Karmaker AK, Rahman MM, Hossain MA, Ahmed MR (2020) Exploration and corrective measures of greenhouse gas emission from fossil fuel power stations for Bangladesh. J Clean Prod 244:118645. https://doi.org/10.1016/j.jclepro.2019.118645

Kwapinski W, Byrne CMP, Kryachko E, Wolfram P, Adley C, Leahy JJ, Novotny EH, Hilary MHB, Hayes B (2010) Biochar from biomass and waste. Waste Biomass Valori 1(2). https://doi.org/10.1007/s12649-010-9024-8

Lipsett M, Hurley S, Ostro B (1997) Air pollution and emergency room visits for asthma in Santa Clara County, California. Environ Health Perspect 105:216–222

Lange LC, Hills CD, Poole AB (1996) Preliminary investigation into the effects of carbonation on cement-solidified hazardous wastes. Environ Sci Technol 30:25–30

Lange LC, Hills CD, Poole AB (1997) The effect of carbonation on the properties of blended and non-blended cement solidified waste forms. J Hazard Mater 52:193–212

Maries A, Hills CD (1986) Improvement in concrete articles. UK Patent 2(192):392

Malik A (2004) Metal bioremediation through growing cells. Environ Int 30:261–278

Lavric ED, Konnov AA, De Ruyck J (2004) Dioxin levels in wood combustion—a review. Biomass Bioenergy 26(115):145

McConnell R, Berhane K, Gilliland F, London SJ, Islam T, Gauderman WJ, Avol E, Margolis HG, Peters JM (2002) Asthma in exercising children exposed to ozone: a cohort study. Lancet 359: 386–391

McKendry P (2002) Energy production from biomass (part 1): overview of anthropogenic biomass. Bioresour Technol 83(1):37–46

Molfino NA, Slutsky AS, Zamel N (1992) The effects of air pollution on allergic bronchial responsiveness. Clin Exp Allergy 22:667–672

Luthe C, Karidio I, Uloth V (1997) Towards control lanadoxins emissions from power boilers fuelled with salt-laden wood waste. Chemosphere 35:557–574

Liu Z, Zhang FS (2009) Removal of lead from water using biochars prepared from hydrothermal liquefaction of biomass. J Hazard Mater 167(1):933–939

Lu H, Zhang W, Yang Y, Huang X, Wang S, Qiu R (2012) Relative distribution of Pb2+ sorption mechanisms by sludge-derived biochar. Water Res 46(3):854–862. https://doi.org/10.1016/j.watres.2011.11.058

Matteson GC, Jenkins B (2007) Food and processing residues in California: resource assessment and potential for power generation. Bioresour Technol 98(16):3098–3105

Mau V (2016) Phases' characteristics of poultry litter hydrothermal carbonization under a range of process parameters. Bioresour Technol 219:632–642

McIlveen-Wright DR, Williams BC, McMullan JT (2001) A reappraisal of wood/red combustion. Bioressour Technol 76:183–190

Mohan D, Pittman CU, Steele PH (2006) Pyrolysis of wood/biomass for bio-oil: a critical review. Energy Fuel 20(3):848–889

Morrison RT, Boyd RN (1983) Organic chemistry. Allyn and Bacon, Inc, Singapore

Mohan D, Pittman Jr CU, Bricka M, Smith F, Yancy B, Mohammad J, Steele PH, Alexandre-Franco MF, Gomez-Serrano V, Gong H (2007) Sorption of arsenic, cadmium, and lead by chars produced from fast pyrolysis of wood and bark during bio-oil production. J Colloid Interface Sci 310(1):57–73

Nguyen TH, Cho H-H, Poster DL, Ball WP (2007) Evidence for a pore-filling mechanism in the adsorption of aromatic hydrocarbons to a natural wood char. Environ Sci Technol 41(4): 1212–1217

Nasir NF, Daud WRW, Kamarudin SK, Yaakob Z (2013) Process system engineering in biodiesel production: a review. Renew Sustain Energy Rev 22:631–639

Pallavi HV, Swamy SS, Kiran BM, Vyshnavi DR, Ashwin CA (2014) Briquetting agricultural wastes as an energy source. J Eng Sci Comp Sci Eng Technol 2:160–172

Pellera FM, Giannis A, Kalderis D, Anastasiadou K, Stegmann R, Wang J-Y, Gidarakos E (2012) Adsorption of Cu (II) ions from aqueous solutions on biochars prepared from agricultural by-products. J Environ Manage 96(1):35–42

Perlack RD (2005) Biomass as feedstock for a bioenergy and bio-products industry: the technical feasibility of a billion-ton annual supply, U.S. Department of Energy & U.S. Department of Agriculture, Oak Ridge, Tenn, USA

Petroeschevsky A, Simpson RW, Thalib L, Rutherford S (2001) Associations between outdoor air pollution and hospital admissions in Brisbane, Australia. Arch Environ Health 56:37–52

Qiu Y et al (2008) Surface characteristics of crop-residue-derived black carbon and lead (II) adsorption. Water Res 42(3):567–574

Reza MT, Mumme J, Ebert A (2015) Characterization of hydrochar obtained from hydrothermal carbonization of wheat straw digestate. Biomass Conv Bioref 5(4):425–435

Regmi P, Moscoso JLG, Kumar S, Cao X, Mao J, Schafran G (2012) Removal of copper and cadmium from aqueous solution using switchgrass biochar produced via hydrothermal carbonization process. J Environ Manage 109:61–69

Ruokojarvi P, Aatamila M, Ruuskanen J (2000) Toxic chlorinated and polyaromatic hydrocarbons in simulated house fires. Chemosphere 41:825–828

Saratale GD, Saratale RG, Banu JR, Chang JS (2019) Biohydrogen production from renewable biomass resources. Chapter 10-biohydrogen (second edition), biomass, biofuels, biochemicals, pp 247–277

Sánchez-García M (2015) Biochar accelerates organic matter degradation and enhances N mineralisation during composting of poultry manure without a 1303 relevant impact on gas emissions. Bioresour Technol 192:272–279

Searle S, Malins C (2013) Availability of cellulosic residues and wastes in the EU. White Paper, The International Council on Clean Transportation. http://theicct.org/sites/default/files/publications/ICCT_EUcellulosic-waste-residues_20131022.pdf

Shima M, Adachi M (2000) Effect of outdoor and indoor nitrogen dioxide on respiratory symptoms in schoolchildren. Int J Epidemiol 29:862–870

Shrestha G, Traina SJ, Swanston CW (2010) Black carbon's properties and role in the environment: a comprehensive review. Sustainability 2(1):294–320

Shang G, Shen G, Wang T, Chen Q (2012) Effectiveness and mechanisms of hydrogen sulfide adsorption by camphor-derived biochar. J Air Waste Manage Assoc 62(8):873–879

Strand V, Svartengren M, Rak S, Barck C, Bylin G (1998) Repeated exposure to an ambient level of NO_2 enhances asthmatic response to a nonsymptomatic allergen dose. Eur Respir J 12:6–12

Suwelack KU (2016) Prediction of gaseous, liquid and solid mass yields from hydrothermal carbonization of biogas digestate by severity parameter. Biomass Conv Bioref 6(2):151–160

Tester JW, Wood DO, Ferrari NA (1991) Energy and the environment in the 21st century. MIT Press, New York

Tolbert PE, Mulholland JA, MacIntosh DL, Xu F, Daniels D, Devine OJ, Carlin BP, Klein M, Dorley M, Butler AJ, Nordenberg DF, Frumkin H, Ryan PB, White MC (2000) Air quality and pediatric emergency room visits for asthma in Atlanta, Georgia, USA. Am J Epidemiol 151: 798–810

Tong X-j, Li J-u, Yuan J-h, Xu R-k (2011) Adsorption of Cu (II) by biochars generated from three crop straws. Chem Eng J 172(2):828–834

Toufiq Reza M, Freitas A, Yang X, Hiibel S, Lin H, Coronella CJ (2016) Hydrothermal carbonization (HTC) of cow manure: carbon and nitrogen distributions in HTC products. Environ Prog Sustain Energy 35(4):1002–1011. https://doi.org/10.1002/ep.12312

Trasande L, Thurston GD (2005) The role of air pollution in asthma and other pediatric morbidities. J Allergy Clin Immunol 115:689–699

Uchimiya M, Lima IM, Klasson KT, Chang S, Wartelle LH, Rodgers JE (2010) Immobilization of heavy metal ions (CuII, CdII, NiII, and PbII) by broiler litter-derived biochars in water and soil. J Agric Food Chem 58(9):5538–5544

UNEP (2011) Resource efficiency: economics and outlook for Asia and the Pacific. (CSIRO Publishing, Canberra

UNEP (2009) Converting waste agricultural biomass into a resource. UNEP Compendium of Technologies. http://www.unep.org/ietc/Portals/136/Publications/Waste%20Management/WasteAgricultural BiomassEST_Compendium.pdf

Vijayaraghavan K, Yun YS (2008) Bacterial biosorbents and biosorption. Biotechnol Adv 26:266–229

Volesky B (2007) Biosorption and me. Water Res 41:4017–4029

Wang X, Xing B (2007) Sorption of organic contaminants by biopolymer-derived chars. Environ Sci Technol 41(24):8342–8348

Wang H, Lin K, Hou Z, Richardson B, Gan J (2010) Sorption of the herbicide terbuthylazine in two New Zealand forest soils amended with biosolids and biochars. J Soil Sediment 10(2):283–289

Whitehead K, Hills CD, MacLeod CL, Carey PJA (2003) Field application of ACT for the remediation of an ex-pyrotechnics site: monitoring data and microstructural studies, WASCON 2003, Proceedings of 5th international conference on the environmental and technical implications of construction with alternative materials, San Sebastian, Spain, 4–6th June, 2003

Woodruff PWR, Wright I, Bullmore E, Brammer M, Howard RJ, Williams SCR, Shapleske J, Rossell S, David A, McGuire P, Murray RM (1997) Auditory hallucinations and the temporal cortical response to speech in schizophrenia: a functional magnetic resonance imaging study. Am J Psychiatry 154:1676–1682

Xu RK, Xiao S-c, Yuan J-h, Zhao A-z (2011) Adsorption of methyl violet from aqueous solutions by the biochars derived from crop residues. Bioresour Technol 102(22):10293–10298

Xue Y, Gao B, Yao Y, Inyang M, Zhang M, Zimmerman AR, Ro KS (2012) Hydrogen peroxide modification enhances the ability of biochar (hydro-char) produced from hydrothermal carbonization of peanut hull to remove 1360 aqueous heavy metals: batch and column tests. Chem Eng J 200:673–680

Yenisoy-Karakaş S, Aygün A, Güneş M, Tahtasakal E (2004) Physical and chemical characteristics of polymer-based spherical activated carbon and its ability to adsorb organics. Carbon 42(3): 477–484

Zoulalian A (2010) Biomass position for renewable energies: main ways of energetic valorization. J Appl Fluid Mech 3:47–54

Sustainable Management of Municipal Solid Waste: Associated Challenges and Mitigation of Environmental Risks

Yuti Desai, Vijay Kumar Srivastava, Geetanjali Kaushik, Rajiv R. Srivastava, Hyunjung Kim, Sadia Ilyas, and Vinay K. Singh

Abstract Nowadays, the generation of waste is increasing in many developing countries like India as a result of the constant growth of industrialization, urbanization, and population. Improper management of municipal solid waste (MSW) not only has detrimental environmental consequences but also poses a risk to public health and raises some other issues, like socioeconomic matters, that are worth discussing. Thus, it is crucial to improve the regular handling of waste collection, segregation of waste, and proper disposal. There are certain technologies for the conversion of waste into energy like gasification, incineration, pyrolysis, and bio-methanation that convert the MSW into a suitable source for renewable energy that is safe and eco-friendly. However, the incineration process generates a significant amount of secondary waste, like bottom ash and fly ash. This chapter describes

Y. Desai
Department of Environmental Studies, Faculty of Science, The Maharaja Sayajirao University of Baroda, Vadodara, Gujarat, India

V. K. Srivastava
The Maharaja Sayajirao University of Baroda, Vadodara, Gujarat, India

G. Kaushik
Department of Civil Engineering, Hi-Tech Institute of Technology, Aurangabad, Maharashtra, India

R. R. Srivastava
Center for Advanced Chemistry, Institute of Research and Development, Duy Tan University, Da Nang, Vietnam

Resource Management, Faculty of Natural Sciences, Duy Tan University, Da Nang, Vietnam

H. Kim · S. Ilyas (✉)
Department of Earth Resources and Environmental Engineering, Hanyang University, Seoul, South Korea

V. K. Singh (✉)
Department of Environmental Studies, Faculty of Science, The Maharaja Sayajirao University of Baroda, Vadodara, Gujarat, India

Department of Chemistry, Faculty of Science, The Maharaja Sayajirao University of Baroda, Vadodara, Gujarat, India

P. Pathak et al. (eds.), *Anthropogenic Environmental Hazards*,
https://doi.org/10.1007/978-3-031-41013-0_10

the status of waste-to-energy plants, the challenges associated with the implementation of waste-to-energy technology, and the possibility of utilizing residue generated after waste incineration.

Keywords Municipal solid waste · Waste to energy · Incineration · Bottom ash · Sustainable waste management

1 Introduction

The worldwide population is estimated to be over 10 billion by 2050 (Ajorloo et al. 2022). This rapid growth of the population contributes to the generation of a large quantity of waste depending on their daily consumption and food habits, lifestyle, and the traditional use-and-throw linear approach (Ilyas et al. 2022). Such waste is termed "municipal solid waste. The World Bank has estimated that global waste generation may reach 3.40 billion metric tons by the end of the year 2050 (World Bank 2018). The generation of waste is increasing in developing countries like India, contributing about 160,000 tons of MSW on a daily basis, which is estimated to reach 436 million tons by 2025 with an annual growth rate of 4.2% (CPCB 2020–2021). The state-wise details of solid waste in India are summarized in Table 1 (CPCB 2022), which reveals that 80,000 tons/day of MSW are treated, 29,4000 tons/day are disposed of by landfilling, and about 51,000 tons/day are unaccounted for. This quantity is huge and alarming with respect to the growing attention paid to the environment (Desai et al. 2023).

Open dumping of MSW can have an adverse effect on the surrounding air quality due to the mixing of particulate matter, bad smells, and gaseous emissions, whereas the soil and water bodies can also get polluted via the bio- or chemical degradation of waste and their seepage through the channels and voids on the grounds (Cremiato et al. 2018). Organic content contained in the waste can attract different bacteria, viruses, and several other pathogens that can cause long lasting or serious illness for living creatures (Awasthi et al. 2019). Additionally, the formation of methane through the enteric fermentation of MSW is considered the third largest source of man-made methane (Annepu 2012), which can greatly contribute to global warming's impact compared to CO_2 emissions and is estimated to have a 21-fold higher value. Due to direct combustion practices, it has been estimated to generate about 1 ton of CO_2eq for each ton of MSW combustion (Botello-Alvareza et al. 2018). Particularly in India, both the direct combustion and landfill disposal of MSW are common practices (Pujara et al. 2019). It creates immense pressure on public organizations and also on local authorities in urban areas for the execution of effective waste management plans in terms of the collection, segregation, treatment, and disposal of MSW by shifting from traditional nonsustainable practices to advanced sustainable management. In this context, Pujara et al. (2019) have designed a framework for an integrated management of MSW (Fig. 1), showing composting and anaerobic digestion for the direct treatment of organic waste, while

Table 1 The state-wise solid waste management status in India (CPCB 2022)

SI. No.	State	Generated waste (TPD)	Collected (TPD)	Treated (TPD)	Landfilled (TPD)
1	Andhra Pradesh	6898	6829	1133	205
2	Arunachal Pradesh	236.5	202.1	Nil	27.5
3	Assam	1199	1091	41.4	0
4	Bihar	4281.3	4013.5	NA	NA
5	Chhattisgarh	1650	1650	1650	0
6	Goa	226.9	218.9	197.5	22
7	Gujarat	10,373.8	10,332	6946	3385.8
8	Haryana	5352.1	5291.4	3123.9	2167.5
9	Karnataka	11,085	10,198	6817	1250
10	Kerala	3543	964.8	2550	NA
11	Madhya Pradesh	8022.5	7235.5	6472	763.5
12	Maharashtra	22,632.7	22,584.4	15,056.1	1355.4
13	Tamil Nadu	13,422	12,844	9430.3	2301
14	Telangana	9965	9965	7530	991
15	Tripura	333.9	317.7	214.1	12.9
16	Uttarakhand	1458.5	1379	779.8	–
17	Uttar Pradesh	14,710	14,292	5520	0
18	West Bengal	13,709	13,356	667.6	202.2
19	Andaman & Nicobar	89	82	75	7
20	Chandigarh	513	513	69	444

NA data not available

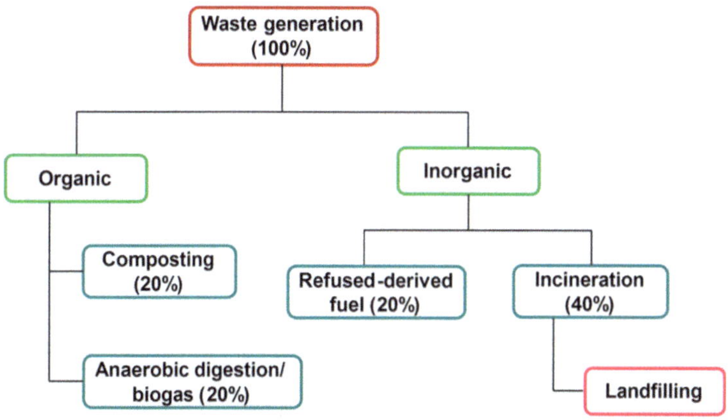

Fig. 1 Framework for an integrated solid waste management system (adopted from Pujara et al. 2019 with copyright permission from Elsevier)

inorganic waste can be treated through refuse-derived fuel and incineration techniques.

2 Waste-to-Energy Concept

The global demands for energy are continuously rising and projected to cross 17 billion metric tons of oil equivalent (toe) by 2035, which will add 29–43 metric tons of CO2 equivalent emissions per year (Chu and Majumdar 2012). Therefore, capturing the energy in terms of fuels and heat by means of the treatments shown in Fig. 1 can lead toward a waste-to-energy (WtE) concept by reducing waste volume and a simultaneous recovery of energy (Istrate et al. 2020). This technique can help reduce environmental issues by reducing fossil fuel consumption and treating waste (Srivastava et al. 2020). They can also be divided into direct processes (involving the production of agro-waste and refuse-derived fuel) and indirect processes (wherein the waste is subjected to burning and heat is captured as a form of energy that involves incineration, composting, gasification, biomethanation, pyrolysis, bioethanol, and landfill gas production) (Malav et al. 2020).

Currently, about 1700 WtE plants worldwide are operating. Among those, Asia Pacific holds 62%, Europe holds 33%, and North America holds 4.5% (UN Environment Annual Report 2019). The primary methods for disposing of waste include open dumping, incineration, landfilling, composting, and recycling on a global scale (as summarized in Table 2). It interestingly reveals that North America has a high prevalence of open dumping, i.e., about 38%, and composting is a less common disposal method. In contrast, incineration is mostly used in East Asia and the Pacific countries (Word Bank 2018). Figure 2 depicts that Japan is the leading country for waste incineration (80.2%), and then Denmark uses 52.5% incineration disposal, followed by Norway (52.3%) and Sweden (51.2%) (World Bank 2019).

Table 2 Worldwide waste treatment and disposal methods (source: "What a Waste 2.0" report of World Bank 2018)

Region	Open dumping	Landfill	Incineration	Composting	Recycling
North America	–	54.3	12	0.4	33.3
South Asia	75	4	–	16	5
Middle East and North America	52.7	34	<1	4	9
Europe and Central Asia	25.6	25.9	17.8	10.7	20
Sub Saharan Africa	69	24	–	<1	6.6
Latin America and Caribbean	26.8	68.5	–	<1	4.5
East Asia and Pacific	18	46	24	2	9
Global average	38.16	36.67	7.83	5.01	12.49

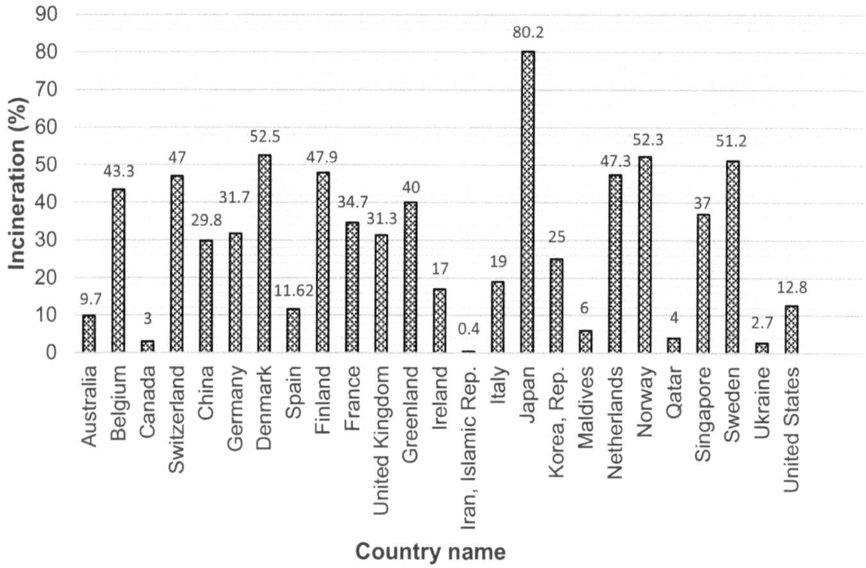

Fig. 2 Country wise waste incineration percentage (source data from World Bank 2019)

2.1 WtE Status-Quo in India

India is the fifth largest generator of MSW with a quantity of 160,000 tons per day (CPCB 2020–2021), albeit it is poorly handled due to a lack of appropriate collection, processing, and disposal (Kumar et al. 2018). In fact, in the last one and a half decades, the generation of MSW has increased four times, as summarized in Fig. 3 (Ministry of New and Renewable Energy 2021). Different approaches for the WtE technologies are attempted in different cities of India. The country has a WtE potential of 2.55 GW and 1.68 GW for the generation of energy from MSW from urban and industrial waste, respectively. About 50.5 GW of energy can be produced by using the MSW generated by cities and towns in the country, which is expected to rise to 1.12 GW by 2031 and 2.8 GW by 2050 (Paulraj et al. 2019). There are 12 WtE facilities running in India, as shown in Fig. 4 (CPCB 2020–2021), where the maximum facilities are in the capital area near the Delhi region. On the other hand, the highest capacity plants are established in the southern Indian state of Andhra Pradesh, and many new projects are in the pipeline, including one in the western Indian state of Gujarat.

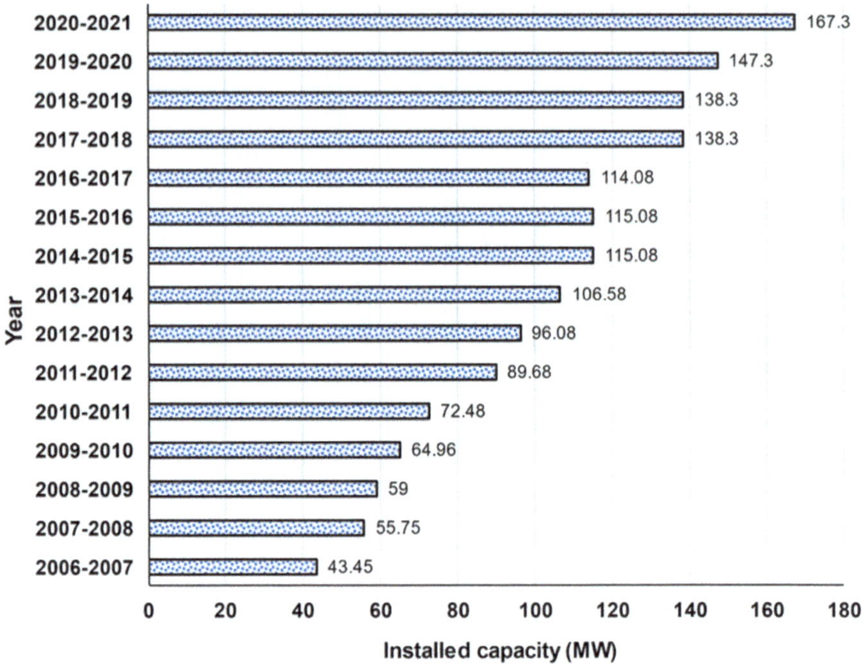

Fig. 3 Waste-to-energy installed capacity in India between the year 2007 and 2018 (Figure is drawn using the sources data from Ministry of New and Renewable Energy 2021)

3 Contributory Techniques in Waste to Energy

3.1 Composting

It is a widespread treatment process for MSW organic waste, which usually accounts for more than 50% of the total weight fraction. Herein, the use of microorganisms, viz., fungi, bacteria, protozoa, and algae, leads to the degradation of organic matter under their growth conditions, wherein organic carbon and nitrogen are consumed as energy sources for the microbial growth. As optimized data suggests, maintaining a pH >4.0 and a temperature between 20 and 40 °C is optimized as a good condition for composting (Tomberlin et al. 2009; Ma et al. 2018). A piece of land, either on-site or off-site, with the proper upholding of aerobic conditions, suitable micro-organisms, optimum temperature, and moisture for microbes' survival is the most important factors for the composting process (Pujara et al. 2019). The thus acquired composts can be primarily used as soil conditioners (Bohacz 2018), resulting in the reduction of MSW volume and turning waste into a source of energy.

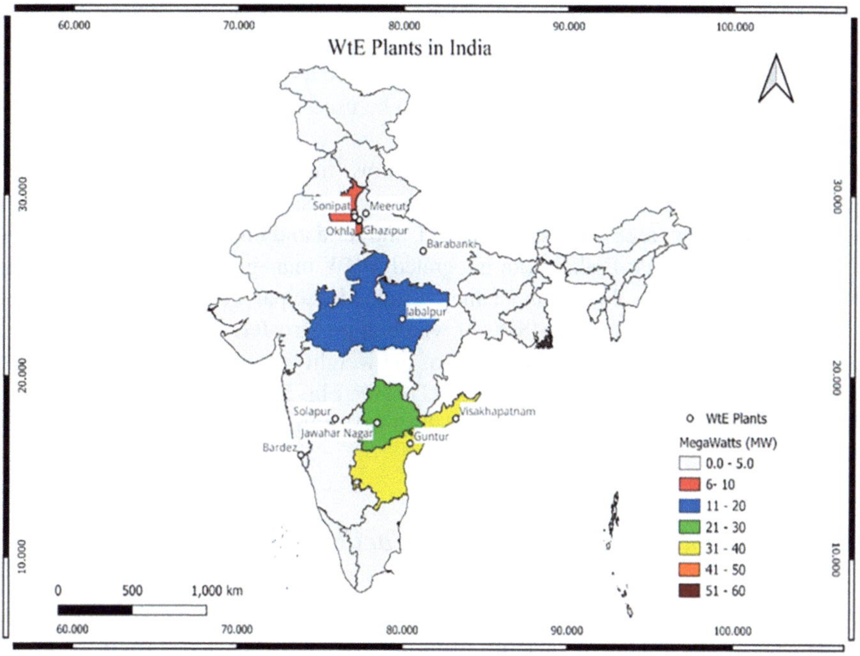

Fig. 4 Installed WtE plants in India (Source data used from Annual Report on Solid Waste Management, CPCB 2020–2021) map is prepared using QGIS version 3.22

3.2 Anaerobic Digestion

Anaerobic digestion, also known as the bio-methanation process, is also applied to the organic waste fraction to be decomposed under anaerobic conditions to generate CO_2 and CH_4 as by-products. On the other side, the residual mass of this process (i.e., digestate) is recycled as manure for soil conditioning. As has been reported by the researchers, a controlled anaerobic digester can produce a fourfold higher quantity of CH_4 by treating 1 ton of MSW for 3 weeks than the landfill treatment of MSW, which usually takes 6–7 years to digest the same quantity with a lower methane value (Ahsan 1999; Pujara et al. 2019). The methane produced can be used as a fuel in industries and for residential purposes as well.

3.3 Refused-Derived Fuels

The refuse-derived fuel (RDF), also termed mechanical biological treatment (MBT), simply utilizes nonbiodegradable combustible waste, viz., papers, plastics, and cardboard. In particular, those cannot be used for recycling purposes but have a higher calorific value (Sharholy et al. 2008). They are used as a feedstock material in a WtE plant to produce the fluffy solid fuels as pallets or cubes for their convenient use as an alternate fuel material in industries. Due to a calorific value in the range of 8–14 MJ/kg, it generates sufficient energy to be used for heating systems in houses and industrial use (Govani et al. 2019).

Although it is not always necessary to follow, the weight distribution of MSW greatly differs depending on factors such as source waste, types of waste, the weather of waste collection, area of waste generated, and food and lifestyle of the inhabitants. For example, the Khat-Prakalp (an integrated MSW management facility located at Nasik in Maharashtra, India) under the Nashik Municipal Corporation is currently collecting about 600 TPD of MSW, of which approximately 48% is dry waste. On average, they are able to convert >41% of the weight fraction into RDF, which is a very good quantity, i.e., about 250 TPD RDF (see Fig. 5), compared to the proposed framework design for the conversion of 20% MWS to RDF (refer to Fig. 1).

3.4 Material Recovery/Recycling Facility

A materials recovery and recycling facility (MRF) is a specialized unit that collects, segregates, and prepares recyclable materials for marketing to end-use manufacturers. Waste is received at the MRF by the collectors, who dump it on the tipping floor, which is then transported through the conveyor belts to the presorting area. Generally, there are two different types: clean and dirty material recovery facilities. The nonrecyclable items are manually sorted out along with 5–45% of the mixed waste (referred to as a dirty MRF) collection, thereby removing the potential hazards and flammable items like spent batteries, propane tanks, and aerosol cans. On the other side, a clean MRF accepts recyclable materials that have been previously separated from MSW at the source end, either by a residential or commercial generator. There are a variety of clean MRFs, like glass, ferrous and nonferrous metals, aluminum cans, PET bottles, newspapers, and magazines. Metals are separated using magnets and eddy current separators. In India, waste segregation at the source is not being practiced properly, which necessitates a centralized material recovery facility to segregate the nonbiodegradable content from the mixed waste (Pujara et al. 2019). Further, it can be directly sent to the authorized recyclers for material recovery. Poor waste management facilities in India contribute 11% of greenhouse gas emissions (Ramachandra et al. 2018).

3.5 High-Temperature Incineration

Incineration is an effective process that deals with the combustion of MSW in the presence of air at a temperature ≥850 °C. It converts the waste material into ash, flue gas, and heat, which are then released into the environment without any further

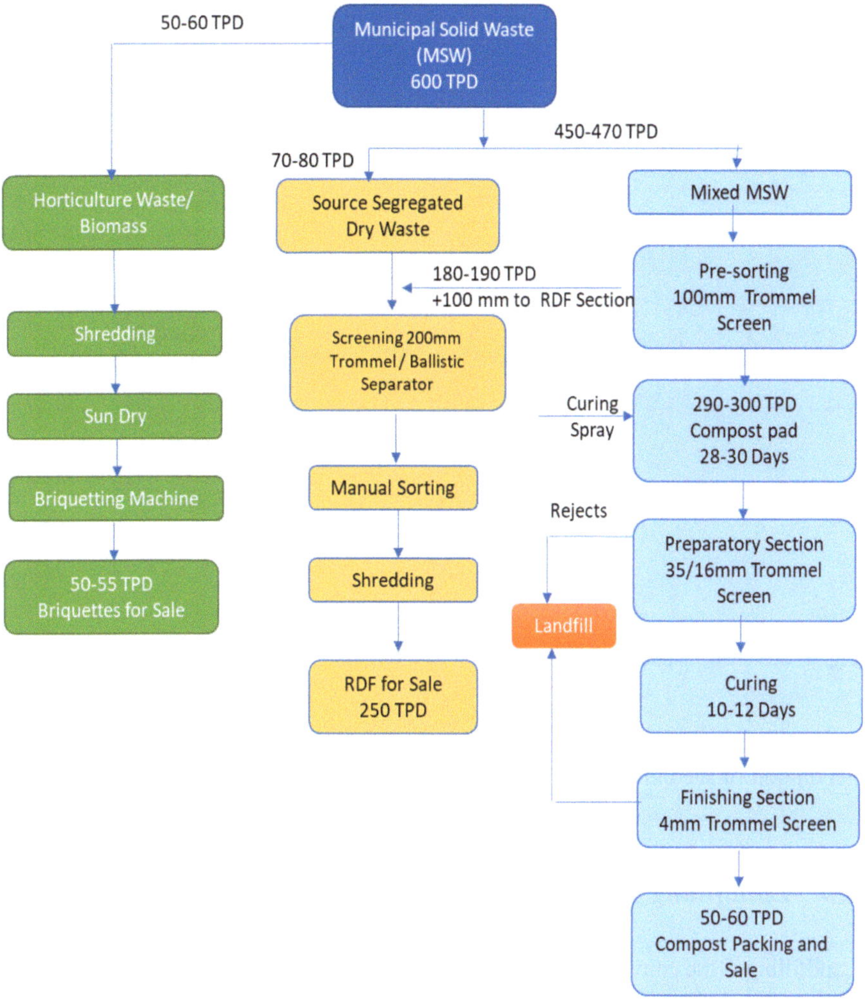

Fig. 5 Integrated solid waste management facility at the Khat-Prakalp, Nasik in the Maharashtra state of India

treatment. A significant amount of heat is generated during waste combustion, which is captured as an energy source. A typical incineration process for MSW is depicted in Fig. 6. However, a variety of incinerators like fixed gates, moving grates, fluidized beds, and rotary kilns have been engineered (Pujara et al. 2019). This process has recently gained attraction due to its simplicity of operation, waste volume reduction of up to 90%, and mass decrease of 70–75% (Desai et al. 2023). It also controls odour and noise and is suitable to operate in various types of weather conditions. In contrast, incineration is not much recommended in the Indian scenario due to the fact that the MSW of India contains high moisture contents (40–60%), which results in a significant reduction of the calorific value found to be 800–1100 kcal/kg (Pujara

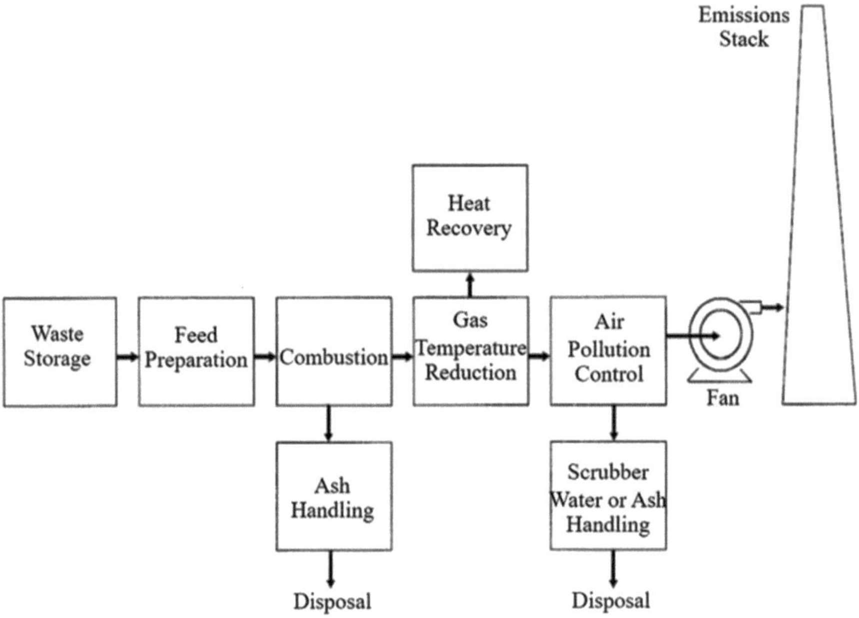

Fig. 6 Schematic diagram of a typical incineration process

et al. 2019). whereas CO_2 emissions during combustion remain an issue with the generation of secondary waste like fly ash and bottom ash (Zhang et al. 2021).

3.6 Landfilling

Landfilling is the conventionally used process for the generated waste volume, and it is also regarded as the final disposal method for the residual waste after treating the MSW. It contains several layers, liners, and covers on top of it. The bottom liners and the top cover of the landfill are considered the most critical components that prevent leachate penetration into the soil. In either case, landfilling is not a sustainable solution. Direct landfilling of MSW produces landfill gases (LFGs) like CH_4 that cause environmental pollution, while the leachate of them can harm human and natural systems through subsurface and groundwater pollution. However, the gases can be treated and further utilized for energy production. The bioreactor landfill is one of the emerging technologies introduced by several cities in India. It utilizes the LFGs for electricity production (Pujara et al. 2023). On the other side, the heavy metals in MSW and residual waste can be harmful to landfills without their proper removal (Desai et al. 2023). A hypothetical sustainability model presented by Pujara et al. (2019) entails maintaining the population on the basis of the carrying capacity of land, wherein waste should be utilized as a resource, which may lead to zero

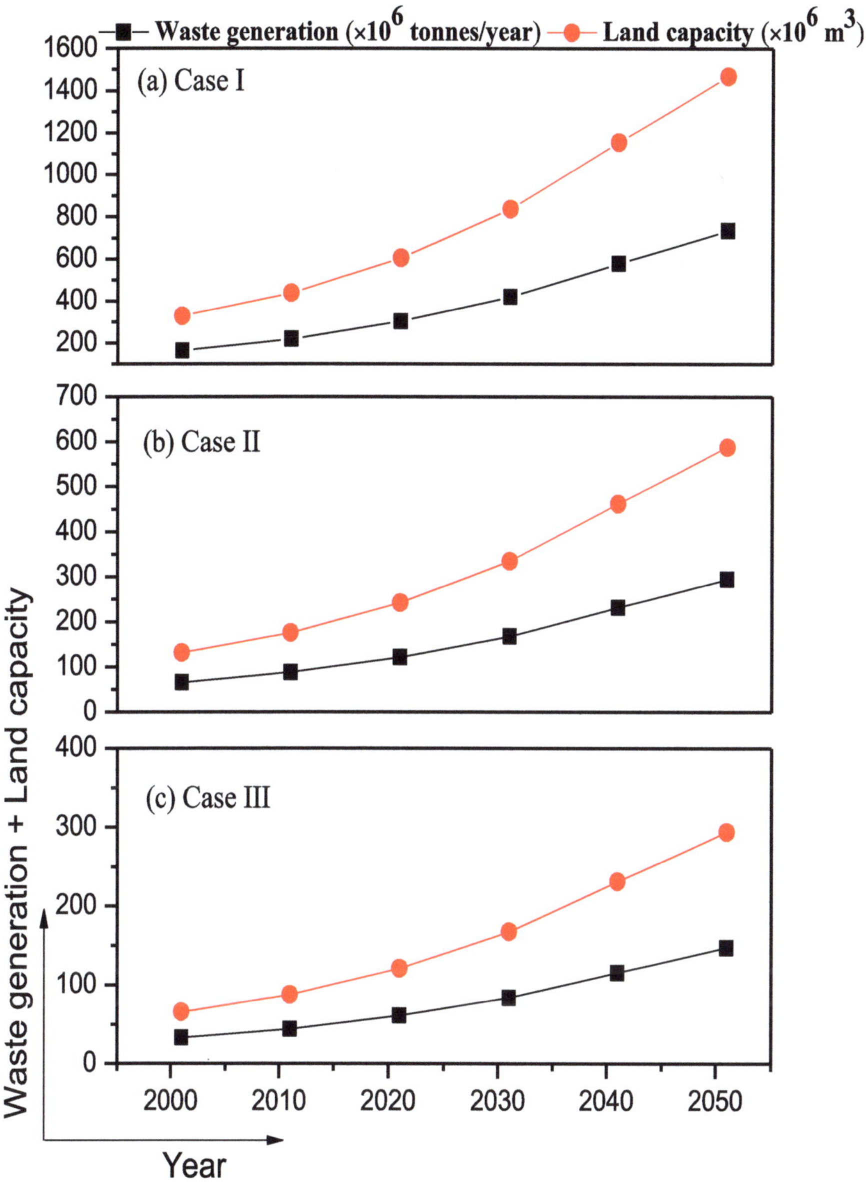

Fig. 7 Different scenario for MSW management with waste generation and land capacity between 2001 and 2050, wherein (**a**) Case-I belongs to the actual scenario which is being observed at present time; (**b**) Case-II represents the controlled and well-managed waste management system and land required in accordance to achieve the sustainable development goals for 2030; and (**c**) Case-III is the most ideal condition where controlled population and controlled waste generation and management system is hypothesized (adopted from Pujara et al. 2019 with copyright permission from Elsevier)

landfilling (refer to Fig. 7). The potential tapping of the WtE program by utilizing the source of MSW can be one of the leading pathways to achieving the bio-energy production goal. Hence, the applicability of life-cycle assessment (LCA) has been found to be helpful for the prior estimation of environmental and economic costs (Gentil et al. 2010), which is discussed below.

4 Assessment of Impacts Through LCA

The LCA is a decision-making tool that is helpful to identify the environmental and economic viability of MSW management (ISO 14040-1997; ISO 14040-2006). The LCA tool involves the following steps: (1) goal and scope; (2) life-cycle inventory; (3) environmental, social, and economic impact assessment; and (4) data interpretation, as depicted in the schematic in Fig. 8 (Koroneos and Nanaki 2012; Kulczycka et al. 2015; Lam et al. 2018). Further, the estimation of environmental loads that result from activities like waste collection, treatment, and disposal can be performed using LCA (Winkler and Bilitewski 2007). Collection, transportation, treatment, landfilling, and composting are thought of as the system boundaries between waste management and the product (Evangelisti et al. 2015; Havukainen et al. 2017). On the other hand, within the unit process system boundaries (like time, place, and functions), the environmental consequences can be assessed (Cleary 2009). Then, the system boundary data can be analyzed by employing software like GaBi, OpenLCA, and SimaPro, which contains several algorithms to compute the LCA (Gentil et al. 2010), thereby using inbuilt libraries and secondary data to give the impact of the process along with the economic and environmental expenses (Winkler and Bilitewski 2007). The traditional practices involved in MSW management have

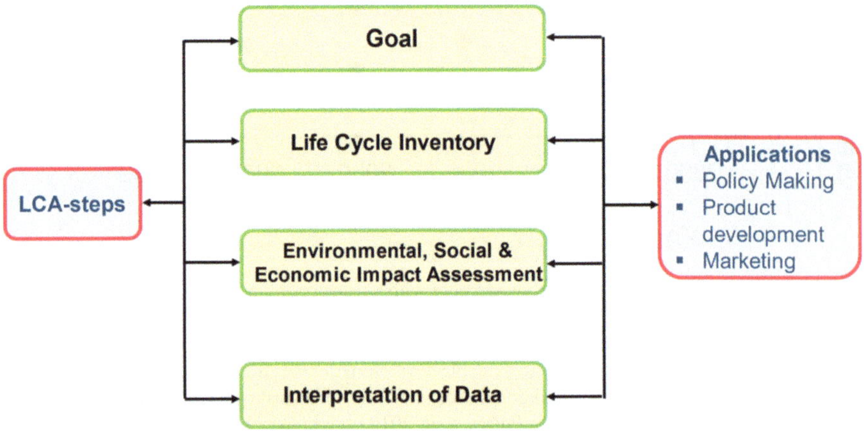

Fig. 8 Key steps of life cycle assessments in MSW management as defined by ISO 14040- (2006) (adopted from Pujara et al. 2019 with copyright permission from Elsevier)

a negative impact on the environment, leading to the abiotic depletion of fossil fuels, global warming potential, acidification potential, ozone layer depletion, resource depletion, human, marine, freshwater, and terrestrial toxicity, and photochemical oxidation potential (Cetinkaya et al. 2018; Fernandez-Nava et al. 2014). Studies revealed, however, that sorting waste and switching raw materials will reduce environmental problems, which will increase environmental and economic benefits (Arena et al. 2003). Further, it is also noted that WtE is suitable for minimizing carbon footprints by shifting the generation of electricity from fossil fuels to waste. Consequently, it prevents uncontrolled methane gas emissions from the landfill site and reduces the potential for global warming (Waste to Energy 2016).

5 Challenges Associated with WtE Process

The development of the WtE industry is facing challenges from a number of political, economic, and technological obstacles, like insufficient funding, a lack of dedicated policies and regulations, and an incomplete data collection and assessment system. Analysis and debate of these constraints are always required for the adoption of future technologies (Khajuria et al. 2010). The failure of WtE technology can be attributed to several factors that include the lack of source segregation, inappropriate techniques for waste collection, missing interests in public participation, the quality of the waste component, litigation concerns, nonfavorable policies, and insufficient financial support in terms of incentives (Istrate et al. 2020). Additionally, the incinerators generate a significant number of gaseous emissions containing pollutants like SO_x, NO_x, dioxins, and furans, whereas residual secondary waste like incinerated fly ash and bottom ash is also challenging to dispose of in the environment (Malav et al. 2020). All these contaminants can have a direct or indirect effect on human health. Henceforth, the local citizens are often not willing to have an incineration facility in their surroundings, which makes it difficult to find a new site to open a new treatment facility. In fact, most of the time, they oppose such initiatives. Such issues can only be solved not only by a solid environmental policy and emissions regulation but also by their strict implementation strategy, which includes a priority on public awareness programs and growing business leaders and investors in the waste management field (Paulraj et al. 2019).

6 Mitigation of the Environmental Issues Related to Heavy Metals in the Secondary Waste of MSW

As per the aforementioned challenges with the WtE treatment model, the generated solid waste is of two types: fly ash (contributing about 20% of the secondary waste generation) and bottom ash (belonging to an 80% weight fraction of the total secondary waste of the incineration process) (Dontriros et al. 2020; Ginés et al. 2009). With the addition of lime during fly ash collection, the acidic nature of it can be neutralized before sending it to landfills, although landfilling is still considered a nonsustainable practice due to the associated heavy metals' contents (Bie et al. 2016). Nevertheless, research suggests that fly ash can be used as a raw material in the creation of building materials like cement, concrete, and glass ceramics (Chen et al. 2020; Fan et al. 2019; Golewski 2018; Zhao et al. 2020).

Numerous studies in this direction reveal that the major application techniques belong to thermal separation, bio- and chemical processing, and electro-chemical processing (Geng et al. 2020a). In a thermal separation, the volatility of metals is achieved at high temperatures that amass in the gas treatment, yielding metal-enriched material for the subsequent metal extraction process. The use of $MgCl_2$/HCl greatly influences the formation of volatile metal chlorides in particular, the base metals like zinc, lead, and copper can effectively volatilize up to 95% by adding $MgCl_2$ at 900 °C, compared to only 87% for lead, 60% for copper, and 40% for zinc without adding the chloride salt. Geng et al. (2020a, b) performed a co-reduction of fly ash with bauxite residue to enhance the reduction efficiency of metals by changing the melting point of the CaO-SiO_2-Al_2O_3 ternary system. They obtained 74% copper, 83% nickel, and 76% iron at the fly ash to bauxite residue mass ratio of 3:7.

On the other hand, the heavy metals' removal is also carried out using different techniques (Weibel et al. 2018; Wang et al. 2021). The acidic fly ash leaching (FLUWA) process and the FLUWA-FLUREC process are among the most feasible processes for metal recycling from fly ash before landfill disposal (Quina et al. 2018). Usually, the acid generated by a wet flue-gas cleaning system is used as the lixiviant medium. In the FLUWA process (refer to Fig. 9), the acidic or alkaline scrub water is slurried with IFA to react in several cascade reactors. After 1 h of leaching, the slurry is filtered to separate the cake and leachate volumes. The leachate undergoes the desired metal separation and recovery process, while the unwanted metals are precipitated as hydroxide by mixing lime in the solution. The cake is sent to landfills for disposal. It is noteworthy to mention that the recovery of heavy metals from IFA does vary depending upon the type of metal and the mineral composition of that metal, as well as factors like waste input for incineration (Desai et al. 2023; Ilyas et al. 2022).

In the case of bottom ash, dry discharge systems can have good results for metal removal and their recovery due to easy screening of particle size fractions, which leads to better treatment potential. The nonagglomerated dry particles reduce the water consumption and transportation costs along with the improved leaching efficacy, although they would need to operate under a closed system to avoid the dust outcome (Böni and Morf 2018; Quicker et al. 2015). A typical treatment facility named MSWI KEZO, Hinwil (Zurich), having a designed capacity of 0.2 million tons per annum of bottom ash, is illustrated in Fig. 10.

As can be seen, the bottom ash is the first magnetic separation of coarse metallic scraps that yield four different particle size fractions: (A) >80 mm, (B) 30–80 mm, (C) 12–30 mm, and (D) <12 mm. The fraction-A is subject to handpicking of large

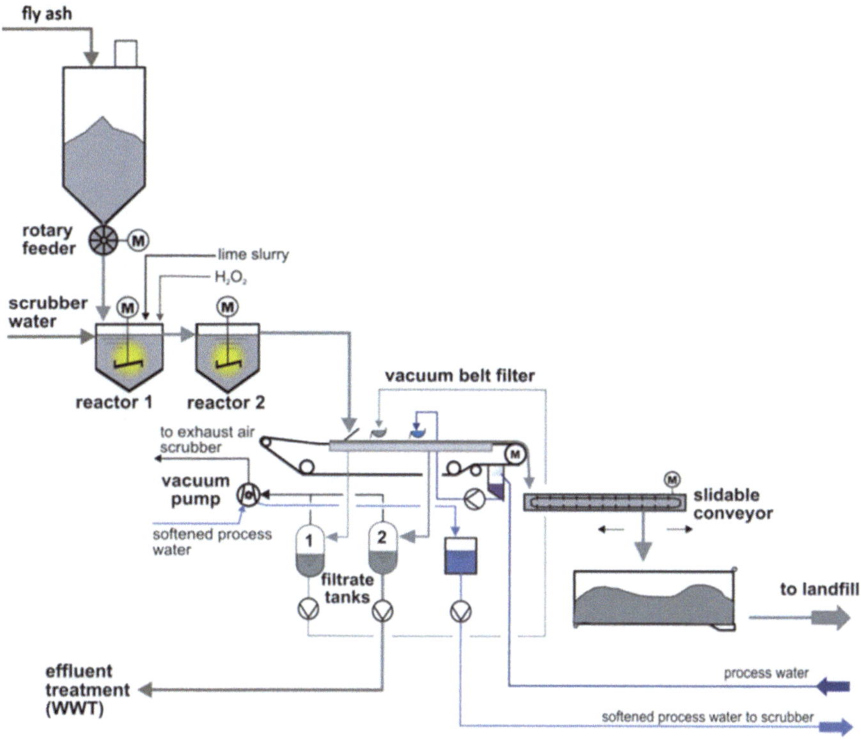

Fig. 9 The flow sheet of the acidic fly ash leaching (FLUWA) process (adopted from Quina et al. 2018, copyright Elsevier)

particle metals to resend for crushing. The fraction-B is subjected to magnetic separation and crushing to obtain particles 30 mm in size. Fraction-C undergoes magnetic separation with a glass separator, two stainless steel separators, ECS, and a crusher to 12 mm. whereas two magnetic separations of fraction-D are followed by two high-frequency ECS to get the light fraction (e.g., Al) separated over the heavier one (e.g., Cu) using the densimetric tables. Consequently, about 10% iron, 4.45% nonferrous metals, and 1.1% glass are obtained.

Several other works have been carried out and reported in the literature for treating the bottom ash and metal recovery. Following a simple water wash, Sun and Yi (2020) achieved the heavy metals' removal with respect to pH swing. A steep increase in pH of the washed liquor (up to 3 h of washing) could be attributed to the dissolution of quicklime and portlandite. Then, the pH started to decline as a result of hydroxyl ion consumption in metal precipitation, forming metal hydroxides (mainly of Pb, Zn, Fe, Cu, and Ni). Therefore, the optimization of a suitable process has been found to be a crucial matter while dealing with heavy metal removal from the bottom ash of MSW incineration.

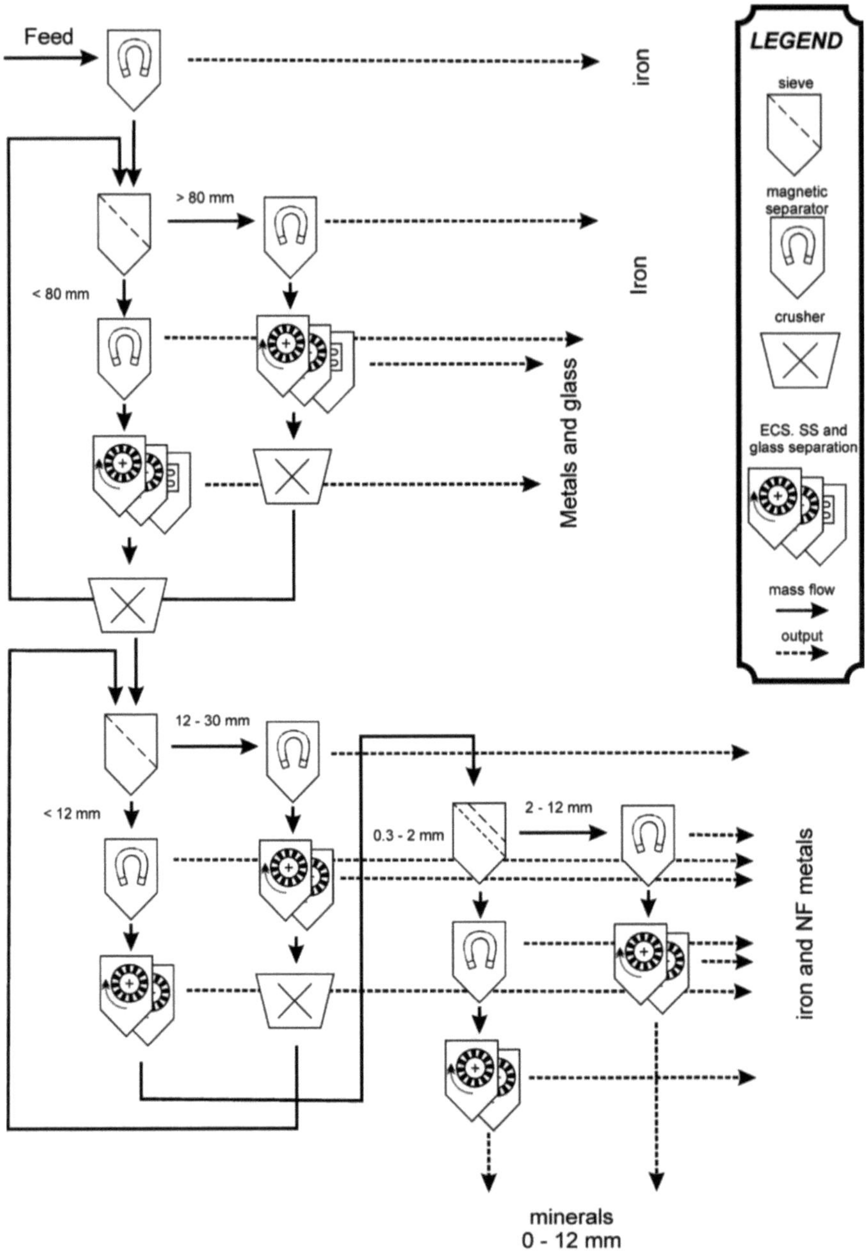

Fig. 10 Scheme of the dry bottom ash treatment plant in Hinwil (CH) (Böni and Morf 2018; adopted with permission from Šyc et al. 2020, copyright Elsevier)

7 Conclusions

Municipal solid waste is an emerging challenge to modern society and sustainable waste management practices. The rapid urbanization and resulting anthropogenic activities cause a significant growth rate in waste generation. Waste to energy (WtE) has been found to be an important treatment practice to follow as a sustainable way to handle waste volume in addition to tapping the advantageous heat and energy from that. The contributory techniques of WtE like composting, anaerobic digestion, refuse-derived fuels, material recovery and recycling facilities, high-temperature incinerations, and landfilling have been discussed along with the life-cycle assessment. At present, the development of the WtE industry is facing challenges from a number of political, economic, and technological obstacles, like insufficient funding, a lack of dedicated policies and regulations, and an incomplete data collection and assessment system. Among other techniques, incineration has been widely used to reduce the mass and volume of MSW, leaving secondary waste to be collected as fly ash and bottom ash, the treatment of which is necessary for heavy metal removal before sending it to final landfilling. The extent of the formation of volatile metal species from fly ash shows great potential for the separate removal of different metals. It has been found that the reducing gas atmosphere increased Zn volatility, although it suppressed Cu volatility and Pb volatility. Whereas metal recovery from bottom ash has remained dependent mainly on dry mechanical processes.

Acknowledgments The authors SI and HK acknowledge their contributions by the Basic Science Research Program through the National Research Foundation of Korea (NRF) funded by the Ministry of Education (Project no. 2023-00243477) and grant funded by the Korea Government (MSIT) (No. 2022R1A5A1032539).

References

Ahsan N (1999) Solid waste management plan for Indian megacities. Indian J Environ Prot 19(2): 90–95

Ajorloo M, Ghodrat M, Scott J, Strezov V (2022) Heavy metals removal/stabilization from municipal solid waste incineration fly ash: a review and recent trends. J Mat Cycles and Waste Manag 24:1693–1717

Annepu RK (2012) Sustainable solid waste management in India, vol 2. Columbia University, New York, pp 1–189

Arena U, Mastellone ML, Perugini F (2003) The environmental performance of alternative solid waste management options: a life cycle assessment study. J Chem Eng 96:207–222

Awasthi MK, Sarsaiya S, Chen H, Wang Q, Wang M, Awasthi SK, Li J, Liu T, Pandey A, Zhang Z (2019) Global status of waste-to-energy technology. Current developments in biotechnology and bioengineering. Elsevier, pp 31–52

Bie R, Chen P, Song X, Ji X (2016) Characteristics of municipal solid waste incineration fly ash with cement solidification treatment. J Energy Inst 89(4):704–712

Bohacz J (2018) Microbial strategies and biochemical activity during lignocellulosic waste composting in relation to the occurring biothermal phases. J Environ Manag 206:1052–1062

Böni D, Morf LS (2018) Thermo-recycling: efficient recovery of valuable materials from dry bottom ash. Removal, Treatment and Utilisation of Waste Incineration Bottom Ash. Thomé-Kozmiensky Verlag GmbH, pp 25–37
Botello-Alvareza JE, Rivas-Garciab P, Fausto-Castro L, Estrada-Baltazar A, Gomez-Gonzalez R (2018) Informal collection, recycling and export of valuable waste as transcendent factor in the municipal solid waste management: a Latin-American reality. J Clean Prod 182:485–495
Cetinkaya AY, Bilgili L, Levent Kuzu S (2018) Life cycle assessment and greenhouse gas emission evaluation from Aksaray solid waste disposal facility. Air Qual Atmos Health 11(5):549–558
Chen W, Wang F, Li Z, Li Q (2020) A comprehensive evaluation of the treatment of lead in MSWI fly ash by the combined cement solidification and phosphate stabilization process. Waste Manag 114:107–114
Chu S, Majumdar A (2012) Opportunities and challenges for a sustainable energy future. Nature 488(7411):294–303
Cleary J (2009) Life cycle assessments of municipal solid waste management systems: a comparative analysis of selected peer-reviewed literature. Environ Int 35–8:1256–1266
CPCB (2020–2021) Annual report on solid waste management. Delhi: CPCB
CPCB (2022) Annual report 2022–21 on implementation of solid waste management rules, 2016. https://cpcb.nic.in/uploads/MSW/MSW_AnnualReport_2020-21.pdf
Cremiato R, Mastellone ML, Tagliaferri C, Zaccariello L, Lettieri P (2018) Environmental impact of municipal solid waste management using life cycle assessment: the effect of anaerobic digestion, materials recovery and secondary fuels production. Renew Energy 124:180–188
Desai Y, Srivastava RR, Srivastava VK, Kaushik G, Singh VK (2023) Hydrometallurgical recovery of critical metals from an incinerated fly ash of municipal solid waste from western India. Geosystem Eng:1. https://doi.org/10.1080/12269328.2023.2201296
Dontriros S, Likitlersuang S, Janjaroen D (2020) Mechanisms of chloride and sulfate removal from municipal-solid-waste-incineration fly ash (MSWI FA): effect of acid-base solutions. Waste Manag 101:44–53
Evangelisti S, Tagliaferri C, Clift R, Lettieri P, Taylor R, Chapman C (2015) Life cycle assessment of conventional and two-stage advanced energy from-waste technologies for municipal solid waste treatment. J Clean Prod 100:212–223
Fan WD, Liu B, Luo X, Yang J, Guo B, Zhang SG (2019) Production of glass–ceramics using Municipal solid waste incineration fly ash. Rare Metals 38(3):245–251
Fernandez-Nava Y, Rio JD, Rodriguez-Iglesias J, Castrillon L, Maranon E (2014) Life cycle assessment (LCA) of different municipal solid waste management options: a case study of asturias (Spain). J Clean Prod 81:178–189
Geng C, Chen C, Shi X, Wu S, Jia Y, Du B, Liu J (2020a) Recovery of metals from municipal solid waste incineration fly ash and red mud via a co-reduction process. Resour Conserv Recycl 154: 104600
Geng C, Liu J, Wu S, Jia Y, Du B, Yu S (2020b) Novel method for comprehensive utilization of MSWI fly ash through co-reduction with red mud to prepare crude alloy and cleaned slag. J Hazard Mater 384:121315
Gentil EC, Damgaard A, Hauschild M, Finnveden G, Eriksson O, Thorneloe S, Kaplan PO, Barlaz M, Muller O, Matsui Y, Ii R, Christensen TH (2010) Models for waste life cycle assessment: review of technical assumptions. Waste Manag 30:2636–2648
Ginés O, Chimenos JM, Vizcarro A, Formosa J, Rosell JR (2009) Combined use of MSWI bottom ash and fly ash as aggregate in concrete formulation: Environmental and mechanical considerations. J Hazard Mater 169(1–3):643–650
Golewski GL (2018) Green concrete composite incorporating fly ash with high strength and fracture toughness. J Clean Prod 172:218–226
Govani J, Patel HT, Chabhadiya K, Jadeja U, Pathak P (2019) Transformation of industrial waste into alternate resource: a critical review. National environmental conference (NEC-2019), IIT Bombay, India

Havukainen J, Zhan M, Dong J, Liikanen M, Deviatkin I, Li X, Horttanainen M (2017) Environmental impact assessment of municipal solid waste management incorporating mechanical treatment of waste and incineration in Hangzhou. China J Clean Prod 141:453–461

Ilyas S, Singh VK, Kim H, Srivastava R (2022) Recovery of metal values by treating the municipal solid waste incineration ashes. Circular economy in municipal solid waste landfilling: biomining & leachate treatment: sustainable solid waste management: waste to wealth. Springer, Cham, pp 253–267

ISO 14040 (1997) Life cycle assessment-principle and guidelines. https://web.stanford.edu/class/cee214/Readings/ISOLCA.pdf

ISO 14040 (2006) Life cycle assessment-principle and guidelines. https://web.stanford.edu/class/cee214/Readings/ISOLCA.pdf

Istrate IR, Iribarren D, Gálvez-Martos JL, Dufour J (2020) Review of life-cycle environmental consequences of waste-to-energy solutions on the municipal solid waste management system. Resour Conserv Recycl 157:104778

Khajuria A, Matsui T, Machimura T, Morioka T (2010) Assessment of the challenge of sustainable recycling of municipal solid waste management in India. Int J Environ Technol Manag 13(2): 171–187

Koroneos CJ, Nanaki EA (2012) Integrated solid waste management and energy production - a life cycle assessment approach: the case study of the city of Thessaloniki. J Clean Prod 27:141–150

Kulczycka J, Lelek L, Lewandowska A, Zarebska J (2015) Life cycle assessment of municipal solid waste management—comparison of results using different LCA models. Pol J Environ Stud 24: 125–140

Kumar S, Negi S, Mandpe A, Singh RV, Hussain A (2018) Rapid composting techniques in Indian context and utilization of black soldier fly for enhanced decomposition of biodegradable wastes—a comprehensive review. J Environ Manage 227:189–199

Lam CM, Yu IKM, Medel F, Tsang DCW, Hsu SC, Poon CS (2018) Life-cycle cost-benefit analysis on sustainable food waste management: the case of Hong Kong International Airport. J Clean Prod 187:751–762

Ma J, Lei Y, Rehman K, Yu Z, Zhang J, Li W, Li Q, Tomberlin JK, Zheng L (2018) Dynamic effects of initial pH of substrate on biological growth and metamorphosis of black soldier fly (Diptera: stratiomyidae). Environ Entomol 47:159–165

Malav LC, Yadav KK, Gupta N, Kumar S, Sharma GK, Krishnan S, Bach QV, Rezania S, Kamyab H, Pham QB, Yadav S, Bhattacharyya S, Yadav VK, Bach QV (2020) A review on municipal solid waste as a renewable source for waste-to-energy project in India: current practices, challenges, and future opportunities. J Clean Prod 277:123227

Ministry of New and Renewable Energy (2021). https://mnre.gov.in/waste-to-energy/current-status

Paulraj CRKJ, Bernard MA, Raju J, Abdulmajid M (2019) Sustainable waste management through waste to energy technologies in India-opportunities and environmental impacts. Int J Renew Energy Res 9(1):309–342

Pujara Y, Pathak P, Sharma A, Govani J (2019) Review on Indian Municipal Solid Waste Management practices for reduction of environmental impacts to achieve sustainable development goals. J Environ Manage 248:109238

Pujara P, Govani J, Patet HT, Pathak P, Mashru D, Ganesh PS (2023) Quantification of environmental impacts associated with municipal solid waste management in Rajkot city, India using life cycle assessment. Environ Adv 12:100364

Quicker P, Stockschläder J, Zayat-Vogel B, Pretz T, Garth A, Koralewska R, Tiffert Y (2015) Wertstoffpotenziale von trocken und nass ausgetragenen Abfallverbrennungsaschen. Mineralische Nebenprodukte Und Abfälle 2:117–135

Quina MJ, Bontempi E, Bogush A, Schlumberger S, Weibel G, Braga R, Funari V, Hyks J, Rasmussen E, Lederer J (2018) Technologies for the management of MSW incineration ashes from gas cleaning: new perspectives on recovery of secondary raw materials and circular economy. Sci Total Environ 635:526–542

Ramachandra TV, Bharath HA, Kulkarni G, Han SS (2018) Municipal solid waste: generation, composition and GHG emissions in Bangalore, India. Renew Sust Energ Rev 82:1122–1136

Sharholy M, Ahmad K, Mahmood G, Trivedi RC (2008) Municipal solid waste management in Indian cities—a review. Waste Manag 28:459–467

Srivastava V, Vaish B, Singh RP, Singh P (2020) An insight to municipal solid waste management of Varanasi city, India, and appraisal of vermicomposting as its efficient management approach. Environ Monit Assess 192:1–23

Sun X, Yi Y (2020) pH evolution during water washing of incineration bottom ash and its effect on removal of heavy metals. Waste Manag 104:213–219

Šyc M, Simon FG, Hykš J, Braga R, Biganzoli L, Costa G, Grosso M (2020) Metal recovery from incineration bottom ash: state-of-the-art and recent developments. J Hazard Mater 393:122433

Tomberlin JK, Adler PH, Myers HM (2009) Development of the black soldier fly (Diptera: stratiomyidae) in relation to temperature. Environ Entomol 38:930–934

UN Environment Annual Report (2019). http://www.unep.org/annualreport/2019/index.php

Wang H, Zhu F, Liu X, Han M, Zhang R (2021) A mini-review of heavy metal recycling technologies for municipal solid waste incineration fly ash. Waste Manag Res 39(9):1135–1148

Waste to Energy (2016) World Energy Council. https://www.worldenergy.org/wpcontent/uploads/2017/03/WEResources_Waste_to_Energy_2016.pdf

Weibel G, Eggenberger U, Kulik DA, Hummel W, Schlumberger S, Klink W, Fisch M, Mäder UK (2018) Extraction of heavy metals from MSWI fly ash using hydrochloric acid and sodium chloride solution. Waste Manag 76:457–471

Winkler J, Bilitewski B (2007) Comparative evaluation of life cycle assessment models for solid waste management. Waste Manag 27:1021–1031

World Bank (2018) What a waste 2.0: a global snapshot of solid waste management to 2050

World Bank (2019). https://datacatalog.worldbank.org/search/dataset/0039597

Zhang Y, Wang L, Chen L, Ma B, Zhang Y, Ni W, Tsang DC (2021) Treatment of municipal solid waste incineration fly ash: State-of-the-art technologies and future perspectives. J Hazard Mater 411:125132

Zhao HL, Liu F, Liu HQ, Wang L, Zhang R, Hao Y (2020) Comparative life cycle assessment of two ceramsite production technologies for reusing municipal solid waste incinerator fly ash in China. Waste Manag 113:447–455

Collaborative Governance and Nonmonetary Compensation Mechanisms for Sustainable Forest Management and Forest Fire Mitigation

Satyam Verma, Ekta Purswani, and Mohammed Latif Khan

Abstract This chapter explores the importance of collaborative governance and nonmonetary compensation mechanisms in the context of sustainable forest management and fire mitigation. It provides a comprehensive analysis of these approaches in the Indian context and other developing nations in the tropics. The chapter discusses the problem of forest fires, the need for collaborative governance, and the role of nonmonetary compensation in achieving sustainable outcomes. It examines various aspects, including community-based initiatives, indigenous knowledge and practices, policy and legal frameworks, case studies, integration of technology, challenges and opportunities, stakeholder engagement and awareness, and synergies with sustainable forest management practices. The chapter concludes by highlighting future directions for enhancing collaborative governance and nonmonetary compensation mechanisms in sustainable forest management and fire mitigation.

Keywords Collaborative governance · Nonmonetary compensation · Sustainable forest management · Fire mitigation · Stakeholder engagement

1 Introduction

Forest fires pose a significant challenge to sustainable forest management, leading to ecological, social, and economic consequences (Lang and Moeini-Meybodi 2021). To effectively address this issue, a multifaceted approach is needed, encompassing collaborative governance and nonmonetary compensation mechanisms. Forest fires have become a pressing apprehension global, including in the Indian context

S. Verma (✉) · E. Purswani
Department of Environmental Sciences, Dr. Harisingh Gour Vishwavidyalaya, Sagar, MP, India
e-mail: satyam.verma@dhsgsu.edu.in

M. L. Khan
Department of Environmental Sciences, Dr. Harisingh Gour Vishwavidyalaya, Sagar, MP, India

Department of Botany, Dr. Harisingh Gour Vishwavidyalaya, Sagar, MP, India

P. Pathak et al. (eds.), *Anthropogenic Environmental Hazards*,
https://doi.org/10.1007/978-3-031-41013-0_11

(Mohanty and Mithal 2022). Uncontrolled fires often result from a combination of natural factors and human activities, such as land-use changes, agricultural practices, and negligence (Leone et al. 2009; Busico et al. 2019). These fires pose severe threats to ecosystems, biodiversity, livelihoods, and the overall well-being of forest-dependent communities (Ray et al. 2023). Addressing forest fires requires comprehensive strategies that go beyond traditional firefighting approaches.

Collaborative governance plays a crucial role in effective forest management and fire mitigation efforts, both in the Indian context and globally. It emphasizes the involvement of multiple stakeholders, including local communities, government agencies, nongovernmental organizations, and indigenous groups, in decision-making processes and the implementation of sustainable practices (Cadman et al. 2023). By promoting collaboration, information sharing, and collective responsibility, collaborative governance fosters holistic and inclusive approaches to address the complex challenges of forest management and fire mitigation (Yami and Mekuria 2022).

In the Indian context, collaborative governance has gained recognition as an essential approach for ensuring sustainable forest management and mitigating the risk of forest fires. India, with its diverse ecosystems and significant forest cover, faces the dual challenge of balancing economic development with ecological conservation (Srivathsa et al. 2023). Collaborative governance offers a framework for addressing these challenges by engaging stakeholders at various levels and integrating their knowledge, perspectives, and expertise.

One notable example of collaborative governance in India is the Joint Forest Management (JFM) approach. Implemented across several states, JFM involves the active participation of local communities in forest management activities. Through this collaborative model, communities are empowered to protect and manage forests, leading to reduced incidences of forest fires. Additionally, JFM has resulted in improved livelihoods for communities by providing them with access to non-timber forest products and ecotourism opportunities (Bhattacharya et al. 2010).

On a global scale, collaborative governance has also proven effective in forest management and fire mitigation. An excellent example is found in Canada, where the Canadian Boreal Forest Agreement (CBFA) was established. The CBFA brought together environmental organizations and the forestry industry to develop sustainable forest management practices that consider ecological, social, and economic aspects. This collaborative approach has contributed to reducing conflicts, promoting conservation measures, and implementing effective fire management strategies (Teitelbaum et al. 2023; Riddell 2014).

Furthermore, in Sweden, the Sami people have played a significant role in collaborative governance practices for forest management. The Sami, an indigenous community, have traditional knowledge and practices that have been integrated into modern forest management strategies. Their involvement and contributions have led to more sustainable forest management approaches and effective fire mitigation measures (Parkatti and Tahvonen 2021). The importance of collaborative governance in forest management and fire mitigation extends beyond specific regions or countries. By bringing together diverse stakeholders and fostering inclusive decision-making processes, collaborative governance ensures that the interests and perspectives of different groups are considered. It enhances the effectiveness and legitimacy of forest management efforts, promotes the conservation of biodiversity, and safeguards ecosystem services (Nikolakis and Hotte 2020).

Collaborative governance is of paramount importance in both the Indian context and globally for sustainable forest management and fire mitigation. It enables the integration of diverse knowledge systems, encourages stakeholder participation, and supports the development of inclusive policies and practices. By embracing collaborative governance, countries can harness the collective wisdom and efforts of various stakeholders, leading to more resilient forests, reduced forest fire risks, and the promotion of sustainable development (Srivathsa et al. 2023). Nonmonetary compensation mechanisms on the other hand play a significant role in promoting sustainable forest management. While financial incentives are crucial, nonmonetary compensation goes beyond monetary values and addresses broader ecological, social, and cultural aspects. These mechanisms recognize and enhance the multiple benefits that forests provide beyond economic gains, fostering a more comprehensive and balanced approach to conservation and management (Tedesco et al. 2023; Lundgren and Morrison-Métois 2016).

1. Recognition of Ecosystem Services: While monetary compensation schemes have been used a lot under the REDD (reducing emissions from deforestation and forest degradation), their effect on alleviating poverty has been insignificant (Lundgren and Morrison-Métois 2016). Another report by Nordic Council of Environmental Ministers (2015) suggests that compensation mechanisms due to their clear framework remain largely ineffective. Nonmonetary compensation mechanisms protect the underlying theme of UN Decade on Ecosystem Restoration which is addressing governance and socioeconomic challenges along with conservation (Santangeli et al. 2016). Forests offer benefits such as carbon sequestration, water regulation, biodiversity conservation, and cultural values. By acknowledging and incorporating these services into management strategies, nonmonetary compensation mechanisms ensure their preservation and sustainable use (Ainsworth et al. 2022).
2. Conservation of Biodiversity: Forests are home to a rich array of plant and animal species. Nonmonetary compensation mechanisms support the conservation and restoration of biodiversity by promoting measures such as protected areas, habitat restoration, and ecological corridors. These mechanisms recognize the intrinsic value of biodiversity and its crucial role in maintaining ecosystem health and resilience (Wood et al. 2019). These mechanisms recognize the intrinsic value of biodiversity and its crucial role in maintaining ecosystem health and resilience (Wood et al. 2019). One of the CBD targets is to phase out the subsidies that amount to biodiversity loss of at least $500 bn annually.
3. Participation and Empowerment: Nonmonetary compensation mechanisms encourage stakeholder participation and empower local communities and indigenous groups. By involving them in decision-making processes, their traditional knowledge, practices, and cultural values can be integrated into forest management strategies. This approach fosters a sense of ownership, strengthens

community resilience, and promotes sustainable practices that align with local contexts. Convention on biological diversity also aims to encourage people's responsible attitudes and participation in forest conservation (Santangeli et al. 2016, Convention on Biological Diversity 2014).

4. Sustainable Livelihoods: Nonmonetary compensation mechanisms contribute to the development of sustainable livelihoods for forest-dependent communities. For instance, they may provide opportunities for community-based forest enterprises, eco-tourism initiatives, or the sustainable harvesting of non-timber forest products. These mechanisms recognize the importance of balancing economic activities with conservation, ensuring the long-term well-being of forest-dependent communities.
5. Social and Cultural Benefits: Forests hold immense cultural, spiritual, and esthetic value for societies. Nonmonetary compensation mechanisms safeguard these intangible benefits by recognizing and protecting cultural heritage, traditional practices, and sacred sites. They contribute to the preservation of cultural diversity and the intergenerational transmission of knowledge and values associated with forests.
6. Environmental Education and Awareness: Nonmonetary compensation mechanisms often include educational programs and awareness campaigns to promote environmental literacy. These initiatives aim to raise awareness about the value of forests, the importance of biodiversity conservation, and sustainable forest management practices. By fostering environmental consciousness, nonmonetary compensation mechanisms empower individuals to become stewards of forest resources (Sattayapanich et al. 2022).
7. Resilience to Climate Change: Forests play a crucial role in climate change mitigation and adaptation. Nonmonetary compensation mechanisms support initiatives that enhance forest resilience to climate change impacts, such as reforestation and forest restoration programs. These mechanisms recognize the role of forests in carbon sequestration, regulating microclimates, and mitigating natural hazards (Wood et al. 2019).

Nonmonetary compensation mechanisms bring essential value to sustainable forest management. They acknowledge the multifaceted benefits of forests, promote stakeholder participation, empower local communities, conserve biodiversity, and enhance cultural and social well-being. By incorporating these mechanisms, forest management strategies can achieve a more balanced and holistic approach, ensuring the long-term sustainability and resilience of forest ecosystems (Walle and Nayak 2020).

2 Collaborative Governance in Forest Management

Collaborative governance has emerged as a vital approach to addressing complex societal challenges and fostering effective decision-making processes (Yami and Mekuria 2022). This section delves into the principles and frameworks that underpin collaborative governance, providing a deeper understanding of its key components and how it can be applied in diverse contexts.

Collaborative governance refers to a collaborative and inclusive process of decision making, in which multiple stakeholders with diverse backgrounds and perspectives come together to jointly address common problems. It recognizes that no single entity or organization can adequately tackle complex issues alone, necessitating the collective expertise, resources, and efforts of different stakeholders.

Several principles form the foundation of collaborative governance:

1. Inclusiveness: Collaborative governance values inclusiveness, ensuring that all relevant stakeholders have the opportunity to participate in decision-making processes. It aims to involve a broad range of voices, including local communities, government agencies, NGOs, academia, and private sector entities. Inclusiveness promotes diverse perspectives, fosters innovation, and strengthens the legitimacy and acceptance of decisions.
2. Shared Purpose: Collaborative governance centers around a shared purpose or common goal that unites stakeholders. This shared purpose serves as a driving force for collaboration, aligning interests and promoting cooperative actions. It requires stakeholders to recognize the interdependence of their goals and to work together towards mutually beneficial outcomes.
3. Trust and Relationships: Building trust among stakeholders is critical in collaborative governance. Trust facilitates open communication, knowledge sharing, and the willingness to explore innovative solutions. Developing strong relationships based on respect, transparency, and shared understanding helps establish a foundation of trust necessary for effective collaboration.
4. Co-creation of Knowledge: Collaborative governance emphasizes the cocreation of knowledge and information among stakeholders. It recognizes that each stakeholder brings unique insights and expertise to the table. By sharing knowledge, combining different perspectives, and engaging in joint learning, stakeholders can collectively generate innovative and informed decisions.
5. Adaptive Management: Collaborative governance embraces adaptive management, recognizing that solutions to complex problems may evolve over time. It encourages learning from experiences, monitoring outcomes, and adjusting strategies as necessary. This adaptive approach allows for flexibility, responsiveness, and continuous improvement in addressing challenges.

Frameworks and models provide guidance for implementing collaborative governance in practice. One widely recognized framework is the "Collaborative Governance Model," which outlines key steps for effective collaboration, including problem identification, stakeholder identification and engagement, goal setting, decision making, and monitoring and evaluation. This model helps structure the collaborative process, ensuring clarity and coherence in collaborative governance initiatives.

Additionally, there are various tools and techniques available to support collaborative governance, such as facilitation, mediation, conflict resolution, and consensus-building processes. These tools help manage disagreements, navigate power dynamics, and foster constructive dialogue among stakeholders. Collaborative governance principles and frameworks have been successfully applied in diverse fields, including natural resource management, environmental policy, urban planning, and public health (Emerson et al. 2012). They enable stakeholders to transcend traditional silos, bridge differences, and develop innovative solutions that reflect the collective wisdom and expertise of those involved.

Understanding collaborative governance principles and frameworks is essential for fostering effective collaboration and decision-making processes. By embracing inclusiveness, shared purpose, trust, cocreation of knowledge, and adaptive management, stakeholders can address complex challenges more holistically and generate sustainable solutions. Applying collaborative governance principles in practice, supported by relevant frameworks and tools, enables stakeholders to navigate complexities, build resilience, and achieve collective outcomes for the benefit of society as a whole.

3 Nonmonetary Compensation Mechanisms for Forest Fire Mitigation

3.1 Introduction to Nonmonetary Compensation Approaches

In the context of forest fire mitigation, nonmonetary compensation mechanisms offer valuable strategies that extend beyond financial incentives. These mechanisms recognize the diverse ecological, social, and cultural values associated with forests and provide alternative means of compensation that contribute to sustainable forest management. This section explores nonmonetary compensation approaches in the Indian context and other developing nations in the tropics, highlighting specific examples from India. Nonmonetary compensation approaches encompass a range of strategies that go beyond monetary rewards. They aim to address the multifaceted dimensions of forest fire mitigation, including ecological restoration, community well-being, and cultural preservation. By recognizing and incorporating these aspects, nonmonetary compensation mechanisms enhance the effectiveness and sustainability of forest fire management strategies.

3.2 Exploring Community-Based Initiatives and Participation in Forest Fire Mitigation

Community-based initiatives play a crucial role in forest fire mitigation, particularly in developing nations in the tropics, including India. These initiatives emphasize the active participation and engagement of local communities in fire management efforts. By involving communities, these approaches empower individuals, foster a sense of ownership, and enhance the resilience of forest ecosystems (Walle and Nayak 2020).

In India, examples of community-based initiatives for forest fire mitigation abound. For instance, the Bishnoi community in Rajasthan has a long-standing tradition of protecting forests, including efforts to prevent and manage forest fires. Their collective efforts and traditional knowledge have contributed significantly to the conservation of forests and the mitigation of fire risks. Furthermore, the Joint Forest Management (JFM) program implemented in various states of India involves the collaboration between local communities and forest departments in forest management activities, including fire prevention and control. Under the JFM model, local communities actively participate in fire line construction, fire patrolling, and early warning systems. These initiatives not only enhance forest fire mitigation but also promote sustainable livelihoods and community empowerment.

3.3 Role of Indigenous Knowledge and Practices in Fire Management

Indigenous communities in India possess valuable traditional knowledge and practices related to fire management. Their indigenous knowledge systems have evolved over generations and offer valuable insights into preventing and managing forest fires. In the Indian context and other developing nations in the tropics, incorporating indigenous knowledge and practices into fire management strategies can significantly enhance their effectiveness.

One example of the integration of indigenous knowledge is the use of controlled or prescribed burning techniques practiced by indigenous communities in the Western Ghats region of India. The Soliga tribe in Karnataka has a deep understanding of forest ecosystems and has been practicing controlled burning for centuries (Madegowda 2009). These controlled burns help maintain biodiversity, regenerate natural vegetation, and prevent the spread of wildfires by reducing fuel loads.

Furthermore, indigenous communities in Northeast India, such as the Mizo and Naga tribes, have traditional practices of shifting cultivation that involve controlled burning as part of their agricultural cycle. These practices not only promote food security but also contribute to fire management by reducing fuel accumulation and minimizing the risk of uncontrolled wildfires. Recognizing and valuing indigenous knowledge systems contributes to the preservation of cultural heritage, promotes social equity, and supports indigenous communities' rights and aspirations. Collaborative efforts that involve indigenous communities in decision-making processes and give them a voice in fire management policies can ensure that their knowledge and practices are respected and effectively integrated into broader forest fire mitigation strategies.

Nonmonetary compensation mechanisms for forest fire mitigation encompass a range of approaches that extend beyond monetary incentives. In the Indian context and other developing nations in the tropics, community-based initiatives and the incorporation of indigenous knowledge and practices play integral roles in effective fire management. These approaches empower local communities, foster sustainable practices, and contribute to the resilience and well-being of forest ecosystems. By integrating nonmonetary compensation mechanisms into forest fire mitigation strategies, a more holistic and inclusive approach can be adopted, promoting sustainable forest management in these regions.

4 Policy and Legal Frameworks for Collaborative Governance and Nonmonetary Compensation

4.1 Overview of Relevant Policies and Laws in India

In India, the implementation of collaborative governance and nonmonetary compensation mechanisms for environmental management, including forest fire mitigation, is supported by various policies and legal frameworks. These frameworks provide a regulatory framework, guidelines, and incentives to promote inclusive decision-making processes, community participation, and the integration of nonmonetary compensation approaches. This section provides an overview of some key policies and laws in India that support collaborative governance and nonmonetary compensation for forest fire mitigation.

1. Forest Rights Act (2006): The Forest Rights Act recognizes the rights of forest-dwelling communities, particularly Scheduled Tribes and other traditional forest dwellers, over forest lands and resources. It empowers communities to participate in the management and conservation of forests and recognizes their traditional knowledge and practices. Under this act, communities have the right to protect and conserve forests, including measures to prevent and control forest fires.
2. National Forest Policy (1988, revised in 2006): The National Forest Policy provides a comprehensive framework for forest management in India. It emphasizes the involvement of local communities, NGOs, and other stakeholders in participatory forest management. The policy encourages collaborative approaches, joint forest management, and the integration of traditional knowledge systems for sustainable forest management, including fire prevention and mitigation.
3. National Disaster Management Act (2005): The National Disaster Management Act establishes a legal framework for managing disasters, including forest fires. It provides guidelines for disaster preparedness, response, and mitigation. Collaborative governance principles are incorporated through the involvement of multiple stakeholders, including local communities, in disaster management planning and implementation.
4. Joint Forest Management (JFM) Guidelines (1990, revised in 2000): The JFM guidelines provide a framework for collaboration between forest departments and local communities in forest management activities. It encourages the participation

of communities in fire prevention, control, and awareness programs. Under the JFM model, communities are actively involved in decision making, planning, and implementation of forest fire mitigation measures.

4.2 Analysis of Institutional Arrangements Supporting Collaborative Governance and Nonmonetary Compensation

Institutional arrangements play a crucial role in supporting collaborative governance and nonmonetary compensation mechanisms for forest fire mitigation in India. These arrangements involve the establishment of institutions, frameworks, and mechanisms that facilitate stakeholder participation, coordination, and the integration of nonmonetary compensation approaches. This section analyzes the institutional arrangements that support collaborative governance and nonmonetary compensation in the Indian context (Sattayapanich et al. 2022).

1. Forest Department: The Forest Department, at both the national and state levels, plays a pivotal role in the implementation of forest fire mitigation strategies. It formulates policies, provides technical expertise, and coordinates efforts to prevent, control, and manage forest fires. The Forest Department often collaborates with local communities, NGOs, and other stakeholders to promote collaborative governance and nonmonetary compensation approaches.
2. Joint Forest Management Committees (JFMCs): JFMCs are community-based institutions established under the JFM model. These committees comprise representatives from local communities and the Forest Department. JFMCs facilitate participatory decision making, planning, and implementation of forest management activities, including fire mitigation. They provide a platform for collaboration, knowledge exchange, and the integration of nonmonetary compensation mechanisms.
3. Community-Based Organizations (CBOs): Community-based organizations, such as self-help groups, community forest management committees, and indigenous community organizations, play a significant role in collaborative governance and nonmonetary compensation for forest fire mitigation. These organizations represent the interests of local communities, facilitate community participation, and ensure the inclusion of traditional knowledge and practices in fire management efforts.
4. Research Institutions and Academia: Research institutions and academia contribute to the development of knowledge, research, and capacity building in collaborative governance and nonmonetary compensation. They conduct studies, provide technical expertise, and generate evidence-based recommendations to inform policy and practice. Collaborative partnerships between research institutions, academia, and implementing agencies enhance the effectiveness of fire mitigation efforts.

India has developed policy and legal frameworks that support collaborative governance and nonmonetary compensation for forest fire mitigation. The Forest Rights Act, National Forest Policy, National Disaster Management Act, and JFM guidelines provide the basis for stakeholder participation, community empowerment, and the integration of nonmonetary compensation approaches. Institutional arrangements, such as the Forest Department, JFMCs, community-based organizations, and research institutions, play essential roles in facilitating collaboration and ensuring the effective implementation of nonmonetary compensation mechanisms. By leveraging these policy and institutional frameworks, India can enhance its efforts in collaborative governance and nonmonetary compensation for sustainable forest management and effective forest fire mitigation.

5 Case Studies: Collaborative Governance and Nonmonetary Compensation in Forest Fire Mitigation

5.1 Examining Successful Case Studies of Collaborative Governance and Nonmonetary Compensation Mechanisms

Case studies provide valuable insights into the practical implementation of collaborative governance and nonmonetary compensation mechanisms in forest fire mitigation. These examples demonstrate the effectiveness of these approaches in diverse contexts and highlight the benefits of stakeholder participation, community engagement, and the integration of nonmonetary incentives. This section examines successful case studies from India and other developing nations in the tropics.

1. Uttarakhand Community-Based Forest Fire Management, India: In Uttarakhand, a state in northern India prone to forest fires, a community-based forest fire management initiative has been implemented. Local communities, in collaboration with the Forest Department and NGOs, actively participate in fire prevention, early detection, and fire suppression activities. The initiative involves the establishment of village-level forest fire management committees and the training of community members in fire management techniques. This approach has led to improved fire response, reduced fire incidents, and increased community ownership of forest resources.
2. Bañados del Este Biosphere Reserve, Uruguay: In the Bañados del Este Biosphere Reserve, a wetland ecosystem in Uruguay, collaborative governance, and nonmonetary compensation mechanisms have been successfully employed for fire mitigation. The reserve is managed through a participatory approach that involves the participation of local communities, NGOs, and government agencies. Local residents contribute to fire prevention activities, such as firebreak

construction and controlled burning. In return, they receive nonmonetary incentives, such as access to sustainable livelihood opportunities and eco-tourism development, which serve as compensation for their active engagement in fire management.

5.2 Lessons Learned and Best Practices from These Case Studies

The examination of successful case studies in collaborative governance and nonmonetary compensation provides valuable lessons and best practices for forest fire mitigation efforts. These insights can inform the design and implementation of future initiatives, promote knowledge sharing, and enhance the effectiveness of collaborative approaches. The following lessons and best practices can be derived from the case studies:

1. Community Engagement and Empowerment: Successful case studies highlight the importance of involving local communities in decision-making processes, planning, and implementation of fire mitigation strategies. Active community engagement builds trust, empowers individuals, and strengthens the sense of ownership and responsibility for forest resources. It is essential to recognize and value local knowledge, practices, and cultural perspectives in fire management efforts.
2. Stakeholder Collaboration and Partnerships: Collaborative governance relies on the collaboration and partnerships between diverse stakeholders, including government agencies, local communities, NGOs, research institutions, and academia. Effective coordination, information sharing, and joint efforts are critical for successful fire mitigation. Partnerships enable the pooling of resources, expertise, and efforts towards a common goal.
3. Nonmonetary Incentives and Compensation: Nonmonetary compensation mechanisms play a crucial role in incentivizing community participation and sustaining long-term engagement in fire management. These incentives can take various forms, such as access to alternative livelihood opportunities, capacity-building programs, infrastructure development, and social recognition. It is important to tailor the incentives to the specific needs and aspirations of local communities to ensure their effectiveness (Nakagawa et al. 2022).
4. Adaptability and Learning: Flexibility and adaptive management are essential in collaborative governance and nonmonetary compensation approaches. Learning from experiences, monitoring and evaluation, and continuous improvement enable the identification of successful strategies, as well as the adjustment of approaches that are not yielding the desired outcomes. Feedback loops and knowledge exchange platforms facilitate the sharing of lessons learned and promote adaptive management (Bilbao et al. 2019).

By integrating these lessons and best practices into forest fire mitigation strategies, policymakers, practitioners, and communities can enhance the effectiveness of collaborative governance and nonmonetary compensation mechanisms. These approaches contribute to sustainable forest management, community resilience, and the reduction of forest fire risks in both the Indian context and other developing nations in the tropics.

6 Integrating Technology for Effective Forest Fire Mitigation

6.1 Role of Technology in Enhancing Collaborative Governance and Nonmonetary Compensation Mechanisms

Technology plays a crucial role in enhancing collaborative governance and nonmonetary compensation mechanisms for effective forest fire mitigation. It provides tools and solutions that improve the efficiency, accuracy, and timeliness of fire detection, prevention, and early warning systems. This section explores the role of technology in supporting collaborative governance and nonmonetary compensation approaches in forest fire mitigation efforts.

1. Remote Sensing and Satellite Imagery: Remote sensing technologies, including satellite imagery and aerial surveys, enable the detection and monitoring of forest fires over large areas. These technologies provide real-time data on fire locations, intensity, and spread, allowing for rapid response and effective resource allocation. Remote sensing also facilitates the identification of high-risk areas, enabling proactive fire prevention measures.
2. Geographic Information Systems (GIS): GIS technology enables the integration and analysis of spatial data, facilitating informed decision making in forest fire management. It allows for the mapping of fire-prone areas, the identification of vulnerable communities, and the visualization of fire risks. GIS-based tools support collaborative governance by providing stakeholders with accessible and interactive platforms for data sharing, planning, and coordination.
3. Sensor Networks and IoT Devices: Sensor networks and Internet of Things (IoT) devices can be deployed in forested areas to detect and monitor fire incidents. These devices, such as temperature and humidity sensors, provide real-time data on environmental conditions that contribute to fire risks. IoT devices can also be integrated with early warning systems, triggering alerts and notifications to relevant stakeholders, including local communities and fire management agencies.

6.2 *Examples of Technological Innovations for Forest Fire Detection, Prevention, and Early Warning Systems*

Technological innovations have led to the development of advanced tools and systems for forest fire detection, prevention, and early warning. These innovations enhance the effectiveness of collaborative governance and nonmonetary compensation mechanisms. The following examples highlight some of the technological advancements in forest fire mitigation:

1. Automated Fire Detection Systems: Automated fire detection systems utilize thermal imaging, infrared sensors, and machine learning algorithms to detect fires in real time. These systems can be integrated with existing surveillance infrastructure or deployed as stand-alone units. Examples include fire detection drones, camera networks, and sensor arrays that can detect smoke, heat, or flame signatures.
2. Mobile Applications for Community Reporting: Mobile applications empower local communities to report fire incidents and provide real-time updates on fire risks. These applications often include features such as geolocation, multimedia reporting, and connectivity with fire management authorities. They facilitate community engagement, early response, and information sharing between stakeholders.
3. Early Warning Systems: Early warning systems utilize weather monitoring, fire behavior modeling, and communication technologies to provide timely alerts and warnings about potential fire outbreaks. These systems can be integrated with sirens, SMS alerts, and mobile applications to disseminate information to communities and relevant authorities. Early warning systems help improve response times and enable proactive measures to mitigate fire risks.
4. Fire Behavior Prediction Models: Fire behavior prediction models utilize complex algorithms and input data, such as weather conditions, topography, and fuel moisture, to forecast the behavior and spread of fires. These models assist fire management agencies in decision making, resource allocation, and planning fire suppression strategies. They contribute to more effective collaboration and allocation of nonmonetary compensation resources.

By integrating technology into collaborative governance and nonmonetary compensation mechanisms, forest fire mitigation efforts can benefit from enhanced monitoring, early detection, and efficient response systems. These technological innovations promote information sharing, stakeholder engagement, and the integration of data-driven approaches in forest fire management, leading to more effective outcomes in both the Indian context and other developing nations in the tropics.

7 Challenges and Opportunities in Implementing Collaborative Governance and Nonmonetary Compensation

7.1 Identifying Challenges and Barriers to Implementation

Implementing collaborative governance and nonmonetary compensation mechanisms for forest fire mitigation presents various challenges and barriers. These obstacles can hinder the effective integration of these approaches into existing systems and practices (Yami and Mekuria 2022). This section examines some of the key challenges and barriers encountered in the implementation of collaborative governance and nonmonetary compensation, both in the Indian context and other developing nations in the tropics.

1. Limited Awareness and Understanding: One of the primary challenges is the limited awareness and understanding of collaborative governance and nonmonetary compensation approaches among stakeholders. Many communities, government agencies, and even nongovernmental organizations may not be familiar with these concepts and their potential benefits. Building awareness and providing education about these approaches are crucial to foster support and engagement.
2. Power Dynamics and Stakeholder Engagement: Collaborative governance requires the active participation and engagement of diverse stakeholders, including local communities, government agencies, and NGOs. However, power dynamics, unequal resource distribution, and conflicting interests can impede effective stakeholder engagement. Overcoming these challenges requires addressing power imbalances, promoting inclusive decision-making processes, and ensuring equitable participation.
3. Limited Capacity and Resources: Implementing collaborative governance and nonmonetary compensation mechanisms often requires adequate capacity and resources. Local communities and government agencies may lack the necessary skills, knowledge, and financial resources to actively participate in fire management efforts. Strengthening capacity through training programs, technical support, and resource mobilization is essential to overcome this challenge.
4. Legal and Policy Constraints: Existing legal and policy frameworks may pose challenges to the implementation of collaborative governance and nonmonetary compensation. Inconsistent regulations, conflicting mandates, and bureaucratic procedures can hinder the integration of these approaches into formal systems. Addressing legal and policy constraints requires policy advocacy, institutional reforms, and the alignment of diverse regulatory frameworks.

7.2 *Exploring Opportunities and Potential Solutions*

Despite the challenges, several opportunities and potential solutions exist to facilitate the implementation of collaborative governance and nonmonetary compensation for forest fire mitigation. By capitalizing on these opportunities, stakeholders can overcome barriers and foster a conducive environment for the adoption of these approaches.

1. Knowledge Sharing and Learning Networks: Establishing knowledge sharing platforms, networks, and partnerships can facilitate the exchange of best practices, lessons learned, and innovative solutions. These platforms can bring together stakeholders from different regions, enabling cross-learning, capacity building, and the adaptation of successful strategies to local contexts.
2. Community Empowerment and Ownership: Empowering local communities and fostering their ownership in forest fire mitigation efforts are crucial. By recognizing the traditional knowledge, practices, and cultural values of communities, collaborative governance can be strengthened. Community-led initiatives, participatory planning processes, and the integration of indigenous practices provide opportunities for community empowerment and active involvement.
3. Synergies with Sustainable Development Goals: Collaborative governance and nonmonetary compensation approaches align with the United Nations' Sustainable Development Goals (SDGs). Leveraging these synergies can mobilize resources, enhance policy support, and promote multistakeholder partnerships. Integration with SDGs such as Goal 15 (Life on Land) and Goal 11 (Sustainable Cities and Communities) opens up opportunities for funding, technical assistance, and global collaboration.
4. Technology and Innovation: Harnessing technological advancements and innovation can address some of the implementation challenges. Integrated information systems, data analytics, and remote sensing technologies can enhance monitoring, early warning systems, and decision-making processes. Embracing digital platforms, mobile applications, and geospatial technologies facilitates stakeholder engagement, information sharing, and collaboration.

Implementing collaborative governance and nonmonetary compensation mechanisms for forest fire mitigation presents both challenges and opportunities. Overcoming barriers requires building awareness, addressing power dynamics, strengthening capacity, and aligning legal and policy frameworks. By capitalizing on opportunities such as knowledge sharing, community empowerment, alignment with sustainable development goals, and leveraging technology, stakeholders can enhance the effectiveness of these approaches. These efforts contribute to sustainable forest management, community resilience, and effective forest fire mitigation in both the Indian context and other developing nations in the tropics.

8 Enhancing Stakeholder Engagement and Awareness

8.1 Strategies for Improving Stakeholder Engagement in Collaborative Governance

Effective stakeholder engagement is vital for the success of collaborative governance in sustainable forest management and fire mitigation. This section explores strategies and approaches to enhance stakeholder engagement, ensuring meaningful participation and inclusive decision-making processes.

1. Participatory Decision making: Emphasizing participatory decision-making processes enables stakeholders to have a voice in shaping forest management and fire mitigation strategies. This approach involves engaging stakeholders from the early stages of planning, fostering open dialogue, and incorporating diverse perspectives. Platforms such as community meetings, workshops, and consultative forums provide opportunities for stakeholder engagement and collaborative decision making.
2. Strengthening Partnerships and Networks: Building strong partnerships and networks among stakeholders is crucial for effective collaborative governance. Collaboration between government agencies, local communities, NGOs, academia, and research institutions fosters information sharing, resource mobilization, and coordinated actions. These partnerships create a conducive environment for joint problem-solving, innovation, and shared responsibilities.
3. Capacity Building and Empowerment: Enhancing the capacity of stakeholders is essential to ensure their active engagement and meaningful contributions. Capacity-building programs can provide training on forest management practices, fire prevention techniques, and collaborative approaches. By equipping stakeholders with the necessary knowledge, skills, and resources, they become empowered to actively participate in decision-making processes and implement sustainable forest management practices.
4. Conflict Resolution and Mediation: Collaborative governance may encounter conflicts and disagreements among stakeholders with differing interests. Implementing conflict resolution mechanisms, such as mediation or negotiation processes, can help find common ground and reach consensus. These mechanisms promote understanding, respect for diverse perspectives, and the resolution of conflicts in a constructive manner.

8.2 Importance of Raising Awareness and Building Capacity for Sustainable Forest Management and Fire Mitigation

Raising awareness and building capacity among stakeholders are fundamental for promoting sustainable forest management and fire mitigation practices. This section highlights the significance of awareness-raising and capacity-building efforts to foster a culture of responsible forest management and fire prevention.

1. Knowledge Dissemination: Creating awareness about the importance of sustainable forest management and fire mitigation is key to engaging stakeholders. Disseminating information through various channels, such as workshops, awareness campaigns, educational programs, and digital platforms, helps stakeholders understand the ecological, economic, and social benefits of sustainable practices. Knowledge dissemination should be targeted at diverse audiences, including local communities, policymakers, academics, and the general public.
2. Education and Training: Building capacity through education and training programs equips stakeholders with the necessary skills and knowledge to implement sustainable forest management and fire mitigation strategies. Training initiatives can focus on fire prevention techniques, ecosystem conservation, indigenous knowledge, and the use of modern technologies. Integrating sustainable forest management principles into formal education systems, vocational training programs, and professional development courses ensures a long-term impact (Banerjee et al. 2019).
3. Demonstration Projects and Pilot Initiatives: Implementing demonstration projects and pilot initiatives can showcase the benefits of sustainable forest management and fire mitigation practices. These projects serve as real-life examples, providing tangible evidence of the positive outcomes that can be achieved through collaborative governance and nonmonetary compensation mechanisms. Demonstrations can include the establishment of community-managed forests, agroforestry models, or successful fire prevention strategies, creating practical learning experiences for stakeholders.
4. Knowledge Exchange and Networking: Facilitating knowledge exchange platforms and networking opportunities allows stakeholders to learn from each other's experiences and share best practices. Conferences, workshops, and online forums provide spaces for stakeholders to share insights, innovations, and challenges. Engaging with international networks and global initiatives fosters cross-cultural learning and promotes the adoption of successful strategies from different contexts.

By implementing strategies for improving stakeholder engagement and raising awareness, stakeholders can actively participate in collaborative governance for sustainable forest management and fire mitigation. Enhancing stakeholder involvement and capacity leads to more inclusive decision-making processes, the implementation of effective forest management practices, and the reduction of forest fire risks in the Indian context and other developing nations in the tropics.

9 Synergies with Sustainable Forest Management Practices

9.1 Linkages Between Collaborative Governance, Nonmonetary Compensation, and Sustainable Forest Management

Collaborative governance and nonmonetary compensation mechanisms are closely intertwined with sustainable forest management practices (Targetti et al. 2021). This section explores the linkages between these concepts, highlighting how they complement and reinforce each other in achieving the goal of sustainable forest management.

1. Participatory Decision making: Collaborative governance emphasizes the active participation of stakeholders in decision-making processes. This participatory approach aligns with the principles of sustainable forest management, which prioritize inclusive and transparent decision making. By involving stakeholders in the planning, implementation, and monitoring of forest management activities, collaborative governance enhances the effectiveness and legitimacy of sustainable forest management practices (Banerjee et al. 2019).
2. Conservation and Restoration of Ecosystems: Sustainable forest management aims to conserve and restore forest ecosystems, promoting their long-term ecological health and resilience. Collaborative governance and nonmonetary compensation mechanisms support these objectives by engaging local communities, indigenous peoples, and other stakeholders in the protection and restoration of forest ecosystems. Through their active involvement, these stakeholders contribute to biodiversity conservation, watershed protection, and the sustainable use of forest resources.
3. Integrated Landscape Approaches: Sustainable forest management recognizes the interconnectedness of ecosystems and landscapes. It emphasizes the need for integrated approaches that consider not only forests but also the broader landscape context. Collaborative governance facilitates multistakeholder engagement and coordination, enabling the integration of forest management with other land uses, such as agriculture, infrastructure development, and biodiversity conservation. This integration ensures the sustainable management of forest resources while considering the social, economic, and environmental dimensions of the landscape.

9.2 Building Resilient Ecosystems Through Integrated Approaches

Building resilient ecosystems is a crucial aspect of sustainable forest management and fire mitigation. Integrated approaches that combine collaborative governance, nonmonetary compensation, and sustainable forest management practices contribute to the development of resilient ecosystems (Nuesiri 2015). This section explores the importance of integrated approaches in building ecosystem resilience.

1. Landscape-Level Fire Management: Integrated approaches recognize that fire management should extend beyond individual forests and encompass the entire landscape. Collaborative governance brings together stakeholders from various sectors to develop landscape-level fire management strategies. These strategies include fire prevention measures, early warning systems, and coordinated response mechanisms. By considering the landscape as a whole, integrated approaches enhance the resilience of ecosystems to fire disturbances.
2. Ecosystem-Based Adaptation: Integrated approaches embrace ecosystem-based adaptation strategies to address the impacts of climate change on forest ecosystems. Collaborative governance facilitates the identification and implementation of nature-based solutions, such as forest restoration, agroforestry practices, and the protection of natural firebreaks. These strategies enhance the adaptive capacity of ecosystems, making them more resilient to climate change-induced challenges, including increased fire risks.
3. Sustainable Livelihoods and Community Well-being: Integrated approaches recognize that sustainable forest management should not only focus on ecological outcomes but also consider the well-being of local communities. Collaborative governance ensures that forest management practices and nonmonetary compensation mechanisms are designed to support sustainable livelihoods, social equity, and community resilience. By integrating social and economic dimensions into forest management, integrated approaches contribute to the overall resilience of ecosystems and the well-being of communities.
4. Monitoring and Evaluation: Integrated approaches emphasize the importance of monitoring and evaluation to assess the effectiveness of collaborative governance, nonmonetary compensation, and sustainable forest management practices. By establishing robust monitoring systems, stakeholders can track progress, identify gaps, and adapt their strategies accordingly. Monitoring and evaluation enable the continuous improvement of integrated approaches, ensuring their long-term sustainability and effectiveness.

By recognizing the linkages between collaborative governance, nonmonetary compensation, and sustainable forest management, and by embracing integrated approaches, stakeholders can build resilient ecosystems (Nuesiri 2015). These approaches contribute to the conservation and restoration of forest ecosystems, the protection of biodiversity, and the reduction of forest fire risks in both the Indian context and other developing nations in the tropics.

10 Conclusions

Throughout this book chapter, we have explored the significance of collaborative governance and nonmonetary compensation mechanisms in the context of sustainable forest management and fire mitigation. We have highlighted the importance of understanding the problem of forest fires, the need for collaborative governance, and the role of nonmonetary compensation in achieving sustainable outcomes.

Key findings include:

1. Collaborative governance brings together stakeholders from diverse backgrounds, fostering inclusive decision-making processes and promoting sustainable forest management practices.
2. Nonmonetary compensation mechanisms, such as knowledge sharing, capacity building, and community-based initiatives, play a vital role in incentivizing and empowering stakeholders to actively engage in forest fire mitigation.
3. Indigenous knowledge and practices contribute to effective fire management strategies, emphasizing the value of incorporating traditional ecological knowledge into modern fire management approaches.
4. Policy and legal frameworks provide a supportive environment for collaborative governance and nonmonetary compensation mechanisms. Analyzing relevant policies and institutional arrangements is crucial for identifying areas of improvement and strengthening the implementation of sustainable forest management practices (Roesch-McNally and Rabotyagov 2016).

Looking ahead, there are several key directions to consider for advancing collaborative governance and nonmonetary compensation mechanisms in the field of sustainable forest management and fire mitigation.

1. Strengthening Partnerships and Networks: There is a need to foster stronger partnerships and networks among stakeholders at various levels, including local communities, government agencies, NGOs, research institutions, and international organizations. Collaboration and knowledge exchange facilitate the sharing of experiences, innovations, and best practices, promoting continuous learning and improvement (Banerjee et al. 2019).
2. Integration of Technology: Technological advancements offer great potential for enhancing collaborative governance and nonmonetary compensation mechanisms. Embracing technologies such as remote sensing, Geographic Information Systems (GIS), and early warning systems can improve forest fire detection, prevention, and response, leading to more effective outcomes in sustainable forest management.
3. Scaling up Successful Case Studies: Building upon successful case studies of collaborative governance and nonmonetary compensation mechanisms is crucial for replication and scaling up. Identifying the key factors that contributed to their success and adapting them to local contexts can pave the way for widespread implementation (Isyaku 2021).

4. Adaptation to Climate Change: As climate change continues to impact forest ecosystems, it is essential to integrate climate change adaptation strategies into collaborative governance and nonmonetary compensation approaches. This involves incorporating climate resilience considerations, addressing increased fire risks, and implementing nature-based solutions that enhance ecosystem resilience.
5. Continuous Learning and Evaluation: Regular monitoring and evaluation of collaborative governance and nonmonetary compensation initiatives are essential for identifying gaps, assessing effectiveness, and making informed decisions. Building robust monitoring systems and engaging in adaptive management practices ensure that approaches remain relevant and responsive to changing needs and challenges.

By embracing these future directions, stakeholders can further advance collaborative governance and nonmonetary compensation mechanisms in sustainable forest management and fire mitigation. This will contribute to the preservation of forest ecosystems, the protection of biodiversity, the well-being of local communities, and the reduction of forest fire risks in both the Indian context and other developing nations in the tropics. In conclusion, collaborative governance and nonmonetary compensation mechanisms offer promising pathways toward sustainable forest management and effective forest fire mitigation. Their integration, along with stakeholder engagement, policy support, and technological innovations, will shape the future of forest management practices, ensuring the preservation of invaluable forest resources for generations to come.

References

Ainsworth D, Collins T, d'Amico F (2022) Nations adopt four goals, 23 targets for 2030 in landmark UN biodiversity agreement. Conv Biol Div 19:2022–2012

Banerjee P, Wang HH, Peterson MJ, Grant WE, Peterson TR (2019) Collaborative modeling and social learning in the context of joint forest management in east Sikkim, India. Front Environ Sci 7:154

Bhattacharya P, Pradhan L, Yadav G (2010) Joint forest management in India: experiences of two decades. Resour Conserv Recycl 54(8):469–480

Bilbao BA, Millán A, Leal A, Méndez C, Vessuri H, Mistry J et al (2019) Transforming fire suppression into an intercultural and participative fire management policy in Canaima National Park, Venezuela. A learning together and Indigenous, academic and institutional knowledge integration process. Biodiversidade Brasileira 9(1):316–316

Busico G, Giuditta E, Kazakis N, Colombani N (2019) A hybrid GIS and AHP approach for modelling actual and future forest fire risk under climate change accounting water resources attenuation role. Sustainability 11(24):7166

Cadman T, Maraseni T, Koju UA, Shrestha A, Karki S (2023) Forest governance in Nepal concerning sustainable community forest management and red panda conservation. Land 12(2):493

Emerson K, Nabatchi T, Balogh S (2012) An integrative framework for collaborative governance. J Public Adm Res Theory 22(1):1–29

Isyaku U (2021) What motivates communities to participate in forest conservation? A study of REDD+ pilot sites in Cross River, Nigeria. For Pol Econ 133:102598
Lang Y, Moeini-Meybodi H (2021) Wildfires—a growing concern for sustainable development
Leone V, Lovreglio R, Martín MP, Martínez J, Vilar L (2009) Human factors of fire occurrence in the Mediterranean. Earth Observ Wildland Fires Mediterranean Ecosyst:149–170
Lundgren H, Morrison-Métois S, (2016). Forests and sustainable forest management. Evaluation evidence on addressing deforestation to reduce CO2 emissions. Evaluation Insights OECD, (11).
Madegowda C (2009) Traditional knowledge and conservation. Econ Pol Wkly:65–69
Mohanty A, Mithal V (2022) Managing forest fires in a changing climate. https://www.ceew.in/sites/default/files/ceew-research-on-states-prone-to-forest-wildfires-india-and-mitigation-methods.pdf
Nakagawa M, Lefebvre M, Stenger A (2022) Long-lasting effects of incentives and social preference: a public goods experiment. PloS One 17(8):e0273014
Nikolakis W, Hotte N (2020) How law shapes collaborative forest governance: a focus on indigenous peoples in Canada and India. Soc Nat Resour 33(1):46–64
Nuesiri EO (2015) Monetary and non-monetary benefits from the Bimbia-Bonadikombo community forest, Cameroon: policy implications relevant for carbon emissions reduction programmes. Comm Dev J 50(4):661–676
Parkatti VP, Tahvonen O (2021) Economics of multifunctional forestry in the Sámi people homeland region. J Environ Econ Manag 110:102542
Ray T, Malasiya D, Verma A, Purswani E, Qureshi A, Khan ML, Verma S (2023) Characterization of spatial–temporal distribution of forest fire in Chhattisgarh, India, using MODIS-based active fire data. Sustainability 15(9):7046
Riddell DJ (2014) From the ground up: the story of the Canadian boreal forest agreement. https://doi.org/10.13140/RG.2.1.2226.0087
Roesch-McNally GE, Rabotyagov SS (2016) Paying for forest ecosystem services: voluntary versus mandatory payments. Environ Manag 57:585–600
Santangeli A, Arroyo B, Dicks LV, Herzon I, Kukkala AS, Sutherland WJ, Moilanen A (2016) Voluntary non-monetary approaches for implementing conservation. Biol Conserv 197:209–214
Sattayapanich T, Janmaimool P, Chontanawat J (2022) Factors affecting community participation in environmental corporate social responsibility projects: evidence from mangrove forest management project. J Open Innov Technol Mark Complex 8(4):209
Srivathsa A, Vasudev D, Nair T, Chakrabarti S, Chanchani P, DeFries R, Deomurari A, Dutta S, Ghose D, Goswami VR, Nayak R, Neelakantan A, Thatte P, Vaidyanathan S, Verma M, Krishnaswamy J, Sankaran M, Ramakrishnan U (2023) Prioritizing India's landscapes for biodiversity, ecosystem services and human well-being. Nat Sustain:1–10
Targetti S, Schaller LL, Kantelhardt J (2021) A fuzzy cognitive mapping approach for the assessment of public-goods governance in agricultural landscapes. Land Use Policy 107:103972
Tedesco, A. M., Brancalion, P.H., Hepburn, M.L.H., Walji, K., Wilson, K. A., Possingham, H. P., ..., Rhodes, J.R. (2023). The role of incentive mechanisms in promoting forest restoration. Philos Trans R Soc B, 378(1867), 20210088
Teitelbaum S, Asselin H, Bissonnette JF, Blouin D (2023) Governance in the boreal forest: what role for local and indigenous communities? In: Boreal forests in the face of climate change: sustainable management. Springer International Publishing, Cham, pp 513–532
Walle Y, Nayak D (2020) How can participatory forest management cooperatives be successful in forest resources conservation? An evidence from Ethiopia. J Sustain For 39(7):655–673
Wood A, Tolera M, Snell M, O'Hara P, Hailu A (2019) Community forest management (CFM) in south-west Ethiopia: maintaining forests, biodiversity and carbon stocks to support wild coffee conservation. Glob Environ Chang 59:101980
Yami M, Mekuria W (2022) Challenges in the governance of community-managed forests in Ethiopia. Sustainability 14(3):1478

MIX
Papier aus verantwortungsvollen Quellen
Paper from responsible sources
FSC® C105338

If you have any concerns about our products,
you can contact us on
ProductSafety@springernature.com

In case Publisher is established outside the EU,
the EU authorized representative is:
Springer Nature Customer Service Center GmbH
Europaplatz 3, 69115 Heidelberg, Germany

Printed by Libri Plureos GmbH
in Hamburg, Germany